HYDROGENATION
with LOW-COST
TRANSITION METALS

HYDROGENATION with LOW-COST TRANSITION METALS

Edited by
JACINTO SA
ANNA SREBOWATA

CRC Press
Taylor & Francis Group
Boca Raton London New York

CRC Press is an imprint of the
Taylor & Francis Group, an **informa** business

CRC Press
Taylor & Francis Group
6000 Broken Sound Parkway NW, Suite 300
Boca Raton, FL 33487-2742

First issued in paperback 2017

© 2016 by Taylor & Francis Group, LLC
CRC Press is an imprint of Taylor & Francis Group, an Informa business

No claim to original U.S. Government works

ISBN-13: 978-1-4987-3053-2 (hbk)
ISBN-13: 978-1-138-74743-2 (pbk)

Visit the Taylor & Francis Web site at
http://www.taylorandfrancis.com

and the CRC Press Web site at
http://www.crcpress.com

Contents

Preface

Hydrogenation with Low-Cost Transition Metals describes recent developments in the preparation of catalysts and their catalytic abilities in chemoselective hydrogenation for the production of fine chemicals and pharmaceutical compounds. In particular, we emphasize the use of low-cost metals (Cu, Ni, Fe, and Ag) that are often present in the form of nanoparticles.

Heterogeneous catalysis plays an essential role in improving industrial processes, but its application to fine chemical production is currently limited. Novel technologies, including nanoscience, provide an attractive means toward increasing their application, which would not only improve process kinetics and economics but also ultimately achieve greener and more sustainable processes.

Herein, we compile the latest developments of heterogeneous hydrogenation with low-cost metals, including reaction mechanism schemes, engineering solutions, and perspective for the field. The book is aimed for undergraduate and graduate students, as well as scientists working in the area. It should be mentioned that the goal of this book is to provide an overview of the field. The examples were chosen in order to cover a larger range of scientific experiments as concisely as possible. Therefore, and in the name of the authors, we would like to apologize for any work that has not been referenced or mentioned. The omission was decided simply on the basis of chapter concision. Finally, the book was written in language that we consider accessible to most science undergraduate students. Technical terms were mentioned only when strictly necessary.

We would like to finish on a personal note and thank the chapter authors for their contributions and all the scientific work that made the execution of this book possible.

Jacinto Sá
Anna Śrębowata

Editors

Jacinto Sá (PhD, physical chemistry) is the group leader of the Nanoleaves and Heterogeneous Catalysis Group at Uppsala University and the Modern Heterogeneous Catalysis (MoHCa) Group leader at the Institute of Physical Chemistry, Polish Academy of Sciences. He received his MSc in chemistry, in the field of analytical chemistry, at the Universidade de Aveiro (Portugal), and did his research project at the Vienna University of Technology, Austria. He was awarded a PhD degree in 2007 by the University of Aberdeen, Scotland in the field of catalysis and surface science. In 2007, he moved to the CenTACat Group at Queens University, Belfast, to start his first postdoctoral fellowship under the guidance of Professor Robbie Burch and Professor Chris Hardacre. During his stay there he was awarded an R&D100 for his involvement in the development of Spaci-MS equipment. In 2010, he moved to Switzerland to start his second postdoctoral fellowship at ETH Zurich and the Paul Scherrer Institute. His research efforts were focused on the adaptation of high-resolution x-ray techniques to the study of catalysts and nanomaterials under working conditions. In 2013, he joined the Laboratory for Ultrafast Spectroscopy at the EPFL, Switzerland, to expand the use of high-resolution x-ray techniques into the ultrafast domain and to take advantage of the newly developed XFEL facilities.

Currently Dr. Sá's research efforts are focused on understanding the elemental steps of catalysis, in particular the ones taking place in artificial photosynthesis and nanocatalytic systems used in the production of fine and pharmaceutical chemicals. Dr. SÆs experience in using accelerator-based light sources to diagnose the mechanisms by which important catalytic processes proceed and, more recently, conventional ultrafast laser sources, makes him one of the most experienced researchers in the world in this area. He has more than 75 publications in scientific international journals and more than 30 oral and 50 poster presentations at scientific congresses.

Jacinto SÆis married to Cristina Paun and they are expecting their first child, a baby boy to be named Lucca V. SÆHe is a member of the Portuguese think tank O Contraditorio, part of the English volunteering group of Red Cross Zurich, and a part-time DJ (DJ Sound it). He enjoys fine art, in particular impressionism and surrealism; traveling; music; cinema; and fine dining.

Anna Śrębowata (PhD, physical chemistry) is a Modern Heterogeneous Catalysis (MoHCa) Group member at the Institute of Physical Chemistry, Polish Academy of Sciences in Warsaw. She received an MSc in chemistry, in the field of physical chemistry, at the Jan Kochanowski University in Kielce, Poland. She was awarded a PhD in 2007 by the Institute of Physical Chemistry, Polish Academy of Sciences, in the field of catalysis on metals. In 2007, she moved to the Laboratoire de ReactivitØde Surface UniversitØPierre et Marie Curie in Paris to start her postdoctoral fellowship under the guidance of Professor GØrald DjØga-Mariadassou. Her research efforts were focused on the selective catalytic reduction of NO_x and NO_x traps in cooperation with French industry.

Currently, Dr. Śrębowata s research efforts are focused on the application of modern nanomaterials based on transition metals as catalysts for environmental protection. The aim of her study is catalytic conversion of chloroorganic compounds into value-added products. From January 2013 to December

2014 she was a vice coordinator of international group of research GDRI Catalyse PAN CNRS. Anna Śrębowata is a member of the Polish Chemical Society and the Polish Club of Catalysis. In her private life she is married with three children. She enjoys dancing and cooking.

Contributors

Damian Giziński
Institute of Physical
   Chemistry, Polish
   Academy of Sciences
Warsaw, Poland

Tomiła Łojewska
Institute of Physical
   Chemistry, Polish
   Academy of Sciences
Warsaw, Poland

Cristina Paun
Uppsala, Sweden

Chapter 1

Introduction to Heterogeneous Hydrogenation and Its Application in the Fine Chemicals Industry

Jacinto Sá

Contents

Catalytic hydrogenations are usually carried out with heterogeneous catalysts using molecular hydrogen as a reducing agent. This is an atom economic transformation and undoubtedly the cleanest hydrogenation method. Alternatively, hydrogen donors such as isopropanol or formic acid can be applied in transfer hydrogenations. This chapter aims to highlight some scientific developments and industrial applications of heterogeneous hydrogenation catalysts.

1.1 Introduction

Catalytic hydrogenation is certainly the most widely applicable method for the reduction of organic compounds and belongs to the most important transformations in the chemical industry. There is no attempt to try to review the entire field since this is a very large endeavor and has been, to a certain extent, done elsewhere, in particular for noble metals [1–6]. Instead, I will highlight some scientific developments and industrial applications of heterogeneous hydrogenation catalysts and set the motif for the following chapters, which focus on the use of nanomaterials of cheap and abundant transition metals. This is a rather new area of research, established to circumvent some of the current limitations with their noble metal catalysts counterparts—namely, low abundance, high cost, and, in some cases, toxicity.

When possible, catalytic hydrogenations are carried out with heterogeneous catalysts using molecular hydrogen as a reducing agent. This is an atom economic transformation and undoubtedly the cleanest hydrogenation method, and it will be the focus of the chapter. Alternatively, hydrogen donors such as isopropanol or formic acid can be applied in transfer hydrogenations. Several important inventions have been accomplished in the last 150 years, such as the application of highly dispersed metals (e.g., nickel) in the hydrogenation of organic compounds [7], selective semihydrogenation of C≡C-bonds in the presence of Pd-Pb/CaCO$_3$ catalysts (Lindlar catalyst) [8,9], and, more recently, asymmetric hydrogenations [10,11], pioneered by Knowles and Noyori for which they received the Nobel Prize in chemistry in 2001.

The earliest known hydrogenation reaction with a man-made catalyst was the addition of hydrogen to oxygen mediated by platinum in the Döbereiner's lamp [12], a device commercialized as early as 1823. The French chemist Paul Sabatier is considered the father of the hydrogenation process. Together with James Joyce (American chemist), he discovered that the introduction of trace amounts of a nickel catalyst accelerated the addition of hydrogen to molecules to gaseous hydrocarbons, known today as the Sabatier process, for which Sabatier shared the 1912 Nobel Prize in chemistry.

Wilhelm Normann was awarded a patent in Germany in 1902 and in Britain in 1903 for the hydrogenation of liquid oils, which was the beginning of what is now a worldwide industry. In 1905, the ammonia synthesis, or Haber-Bosch process, was described as a process that involves hydrogenation of nitrogen with an iron catalyst [13], for which Haber received the 1920 Nobel Prize in chemistry, and Ertl the 2008 Nobel Prize in chemistry for the reaction mechanism [14–17]. The mechanism is illustrated in Figure 1.1 [18]. A few decades later, the Fischer-Tropsch process was reported, in which carbon monoxide is hydrogenated to liquid fuels [19], which is the basis of gas to liquid technology.

In 1922, Voorhees and Adams described an apparatus for performing hydrogenation under pressures above 1 atmosphere. The Parr shaker, commercialized in 1926 for the first time, allowed hydrogenations to be carried out at elevated pressures and temperatures. In 1924, the Raney nickel fine powder catalyst was developed and is still widely used in hydrogenation reactions such as the conversion of nitriles to amines or the production of margarine [20,21].

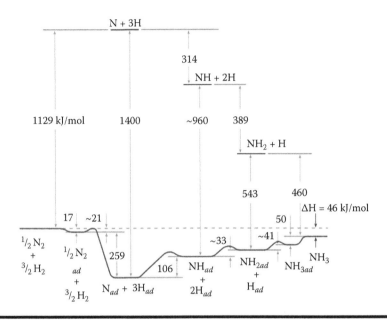

Figure 1.1 **Energy diagram of ammonia synthesis catalyzed by an iron catalyst. (Reproduced from G. Ertl, *Catal. Rev. Sci. Eng.* 21 (1980) 201–223. With permission.)**

Currently, catalytic hydrogenation is employed in a plethora of industrial processes, which often rely on heterogeneous catalysts [22]. In the petrochemical industry, hydrogenation is used to convert alkenes and aromatics into paraffins and naphthenes. Hydrogenation is often used to convert more oxidized compounds containing oxygen and nitrogen, such as aldehydes, imines, and nitriles, into the corresponding saturated compounds (i.e., alcohols and amines). For example, xylitol (sweetener) is manufactured by the hydrogenation of sugar xylose [23]. In respect to nitrile hydrogenation, an example is the synthesis of isophorone diamine, a precursor to the polyurethane monomer isophorone diisocyanate, which is produced from isophorone nitrile by tandem nitrile hydrogenation/reductive amination by ammonia. In food chemistry, the largest scale application is in processing vegetable oils. Typical vegetable oils are derived from polyunsaturated fatty acids (containing more than one carbon–carbon double bonds). Their partial hydrogenation reduces most, but not all, of these carbon–carbon double bonds. The degree of hydrogenation is controlled by restricting the amount of hydrogen, reaction temperature, time, and catalyst [24].

Hydrogenations are strongly exothermic reactions. For example, in the hydrogenation of vegetable oils and fatty acids, the heat released is about 25 kcal per mole (105 kJ/mol), sufficient to raise the temperature of the oil by 1.6°C–1.7°C per iodine number drop. Hydrogenation reactions involving dissociation of the hydrogen molecules on the metal surface have relatively high reaction probability on many surface sites: top site, bridge site, and step sites. For this reason, hydrogen activation is considered structure insensitive. However, the reaction is structure sensitive with respect to the hydrocarbon hydrogenation. For example, Crespo-Quesada et al. [25] reported that semihydrogenation of alkynes occurred preferentially on terraces, whereas further hydrogenation to the alkane occurred over edges (Figure 1.2).

Similarly, Schmidt et al. [26] revealed that in ethyl pyruvate enantioselective hydrogenation, the Pt(111)/Pt(100) ratio controlled both rate and enantiomeric excess. Platinum, palladium, rhodium, and ruthenium are highly active hydrogenation catalysts, operating at low temperatures and low H_2 pressures (Figure 1.3).

The highlighted examples demonstrate the importance of reagents' adsorption, which constitutes the first step of any catalytic reaction. The interaction between reactants and an active site is governed by the electronic structure of the catalyst, which determines adsorption strength and geometry [27–31] (i.e., modification of the electronic structure of the active site alters reactant adsorption parameters). Having information about particle shape is often not sufficient to establish a structure–performance relationship. Take, for example, the α,β-unsaturated aldehyde and ketone hydrogenation on 5 wt% Pt/OMS-2 (Pt average particle diameter 2 nm supported on cryptomelane manganese oxide octahedral

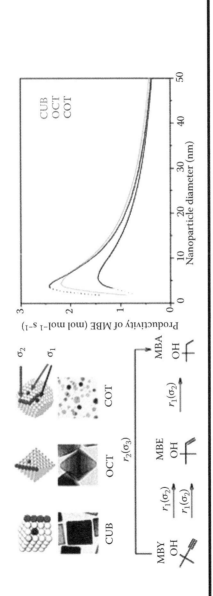

Figure 1.2   Example of a structure sensitivity hydrogenation. The presented case relates to alkynol hydrogenation on shape- and size-controlled palladium nanocrystals. (Reproduced from M. Crespo-Quesada, A. Yarulin, M. Jin, Y. Xia, L. Kiwi-Minsker, *J. Am. Chem. Soc.* 133 (2011) 12787–12794. With permission.)

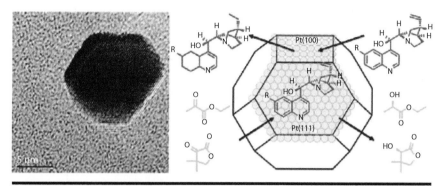

Figure 1.3 Example of a shape-selective enantioselective hydrogenation reaction on Pt nanoparticles. (Reproduced from E. Schmidt, A. Vargas, T. Mallat, A. Baiker, *J. Am. Chem. Soc.* 131 (2009) 12358–12367. With permission.)

molecular sieve). In the case of ketoisophorone (KIP) hydrogenation, 98% selectivity for the selective reduction of the C=C bond occurred, resulting in the production of (6R)-2,2,6-trimethylcyclohexane-1,4-dione (levodione) at 100% conversion. In the case of cinnamaldehyde (CIN) hydogenation, there was 80% selectivity for the reduction of the C=O bond forming cinnamyl alcohol at 100% conversion. The selectivity in KIP hydrogenation contrasted with the expected behavior for Pt, which preferentially hydrogenates the C=O versus C=C bond. The latter selectivity was observed when the reaction was performed over 5 wt% Pt/Al_2O_3 under the same reaction conditions, resulting in 91% selectivity toward 4-hydroxyisophorone at 70% conversion with only 5% selectivity to levodione [32].

The results described were attained in an autoclave reactor operating at 373 K and 10 bar pressure. These conditions are quite challenging for any spectroscopy; thus one needs advanced spectroscopic tools to unveil what is happening. One such technique is photon-in/photon-out high-resolution x-ray spectroscopy because it combines high penetration depth of the probe beam with chemical specificity of the spectroscopy. We performed an in

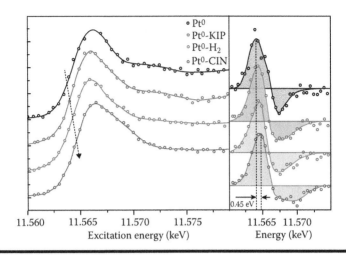

Figure 1.4 Pt L$_3$ edge HERFD-XAS spectra (left) and first derivative (right) of Pt/OMS-2 interacting with H$_2$, KIP and CIN. The spectra were recorded in situ in 10 mol% methanol in water at 373 K and 10 bar pressure. The lines are from a least squares fitting procedure. (Reproduced from H. G. Manyar, R. Morgan, K. Morgan, B. Yang, P. Hu, J. Szlachetko, J. Sá, C. Hardacre, *Catal. Sci. Technol.* 3 (2013) 1497. With permission.)

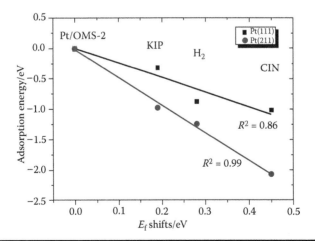

Figure 1.5 Correlation between the experimentally measured shifts in the HERFD-XAS spectra (E_f) and the calculated adsorption energies via DFT for the Pt(111) (black) and Pt(211) (red) surfaces. (Reproduced from H. G. Manyar, R. Morgan, K. Morgan, B. Yang, P. Hu, J. Szlachetko, J. Sá, C. Hardacre, *Catal. Sci. Technol.* 3 (2013) 1497. With permission.)

situ high-energy resolution fluorescence detection—x-ray absorption spectroscopy (HERFD-XAS) study at 373 K and 10 bar pressure (catalyst working conditions) to attain information about the adsorption process, which we combined with density functional theory (DFT) [33]. This was possible due to the development of a home-made cell comprising a stainless steel autoclave reactor with a window made of a polyether ether ketone (PEEK) insert [34]. The resulting shifts in the Pt Fermi energy, measured at Pt L_3-edge, due to adsorption of molecules are depicted in Figures 1.4 and 1.5 and summarized in Table 1.1.

In all cases, molecular adsorption led to a positive shift in the Fermi level energy. This is consistent with what is expected from the Nørskov d-band model since HERFD-XAS at the Pt_3-edge probes these states directly. DFT calculations corroborate the idea that larger Fermi level shifts equate to stronger adsorption energy and that molecules adsorb preferentially on defect surfaces (Figure 1.6), which helped to

Table 1.1   Shifts of Fermi Energy Level (E_f) Position Estimated from Position of Pt LIII Edge Inflection Point for Pt/OMS-2 before and after Adsorption of CIN, KIP, and H_2 Compared with Pt(acac)$_2$ and PtO$_2$

Sample	E_f Shifts Relative to Pt/OMS-2 (eV)	Error (eV)
Pt/OMS-2	0.00	0.13
Pt/OMS-2 + KIP	0.19	0.09
Pt/OMS-2 + H$_2$	0.28	0.08
Pt/OMS-2 + CIN	0.45	0.09
Pt/(acac)$_2$	1.20	0.01
PtO$_2$	2.23	0.20

Source: Reproduced from elsewhere with permission. H. G. Manyar, R. Morgan, K. Morgan, B. Yang, P. Hu, J. Szlachetko, J. Sá, C. Hardacre, *Catal. Sci. Technol.* 3 (2013) 1497.

Note: The errors are determined from the fitting procedure.

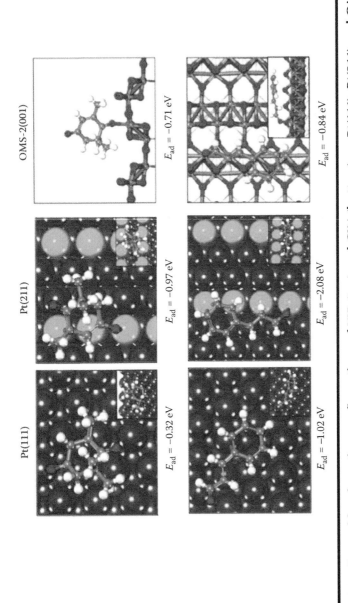

Figure 1.6 Favorable adsorption configurations of KIP (top) and CIN (bottom) on Pt(111), Pt(211) and OMS-2(001). The blue, purple, gray, red, and white balls denote the platinum, manganese, carbon, oxygen, and hydrogen atoms, respectively, and this notation is used throughout the chapter. Those Pt atoms at the edge site of the Pt(211) surface have been highlighted in green. Below each figure panel are the adsorption energies calculated by DFT. (Reproduced from H. G. Manyar, R. Morgan, K. Morgan, B. Yang, P. Hu, J. Szlachetko, J. Sá, C. Hardacre, *Catal. Sci. Technol.* 3 (2013) 1497. With permission.)

understand the difference in reactivity in both systems. DFT calculations also provided clues for adsorption geometry of the molecules on the catalyst, which helped to unveil the reasons for different selectivity (Figure 1.7).

A pertinent question that often occupies scientists is how to modify the metal in order for it to do what it naturally does not want to do. One way is to add a second metal, which can modify the parent catalyst either electronically or geometrically, or both [35]. However, this normally leads only to an enhancement, not a complete switch in selectivity. Postmodification of the parent catalyst can force the molecules to adsorb in a particular mode that switches the catalytic

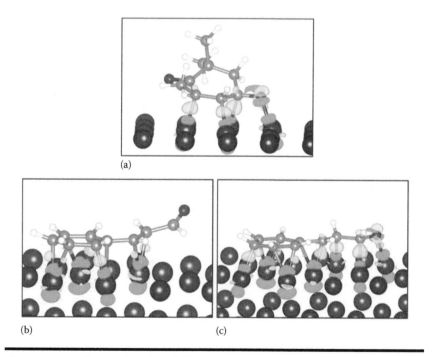

(a)

(b)   (c)

Figure 1.7   Charge density difference analyses on the optimized adsorbed (a) KIP and (b) the most stable and (c) the next most stable states for CIN on the Pt(111) surface. The yellow and green contours show the electron charge accumulation and electron charge depletion induced by the surface binding, respectively. (Reproduced from H. G. Manyar, R. Morgan, K. Morgan, B. Yang, P. Hu, J. Szlachetko, J. Sá, C. Hardacre, *Catal. Sci. Technol.* **3 (2013) 1497. With permission.)**

behavior of the metal [36]. For example, in 4-nitrostyrene hydrogenation, platinum preferentially hydrogenates C=C, leading to the formation of 4-ethylnitrobenzene, although the desired product is 4-aminostyrene that relates to hydrogenation of $-NO_2$ functionality. Makosch et al. [37] demonstrated that product selectivity of Pt/TiO_2 could be switched so that only 4-aminostyrene was produced. This was achieved via postmodification of the parent platinum catalyst with an organic thiol (Figure 1.8). The catalyst was found to be stable at least for three cycles.

1.2 Hydrogenation in the Fine Chemicals Industry

Heterogeneous catalytic hydrogenations are important reactions with a plethora of industrial applications in the production of pharmaceuticals, agrochemicals, fine chemicals, flavors, fragrances, and dietary supplements. The reactions are usually very selective and easy to work up, the catalyst can often be recovered and recycled, and the processes are atom efficient. It comes as no surprise that somewhere between 10% and 20% of the reactions used to produce chemicals today are catalytic hydrogenations [38]. Despite the importance of the technique and mainly because of its multidisciplinary nature, development chemists and engineers have a hard time finding training on this highly specialized subject. The properties required for a heterogeneous catalyst are

■ High activity
■ High selectivity
■ Fast filtration rate
■ Recycle capability

Activity and selectivity are greatly dependent on the choice of metal. This influences the strength of adsorption

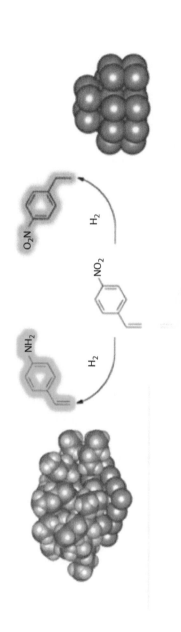

Figure 1.8 Schematic representation of an organic thiol modification of Pt/TiO₂ and the effect on catalyst selectivity. (Reproduced from M. Makosch, W.-I. Lin, V. Bumbálek, J. Sá, J. W. Medlin, K. Hungerbühler, J. A. van Bokhoven, *ACS Catal.* 2 (2012) 2079–2081. With permission.)

of reactants, the rate of desorption of reaction products, and the rate of chemical transformations. One example is the hydrogenation of *cis*-jasmone, which yields different product selectivity depending on whether you use copper or palladium as a parent catalyst (Figure 1.9). The metals most frequently used in heterogeneous catalytic hydrogenation are palladium, platinum, rhodium, nickel, cobalt, and ruthenium. More rarely, iridium, copper, and rhenium find applications.

In the following sections, industrial hydrogenations that are implemented or close to implementation in the pharmaceutical and fine chemical industries are highlighted. In these industries many chemical conversions require stoichiometric amounts of reagents and thus generate large amounts of waste [39,40]. This is in contrast to the production of bulk chemicals, which mostly relies on catalysis. This difference can be explained by the higher complexity of pharmaceuticals and fine chemicals, which makes catalysis more demanding and process development more expensive.

Take, for example, the vitamin industry, which is commonly seen as a typical fine chemical with production volumes of about 100 to 10,000 tons per year [41]. However, some vitamins are produced in much larger quantities that can be placed in the class of bulk chemicals. Typically, these compounds have been produced industrially for decades in

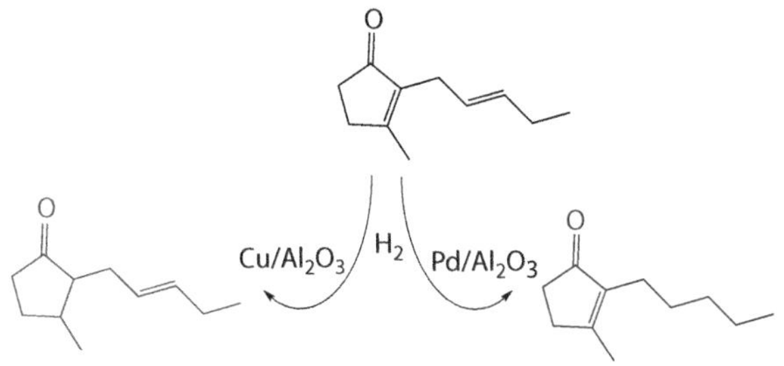

Figure 1.9 Divergent outcomes in the hydrogenation of *cis*-jasmone when catalyzed by copper or palladium.

multistep syntheses with high overall yields. The application of catalytic methods in the highly competitive field of vitamins has increased significantly in recent years because of price pressure on these products but also by the necessity to reduce waste, use less toxic reagents and solvents, improve energy efficiency, recycle catalysts and reagents, and combine unit operations to reduce costs and achieve more sustainable processes.

1.2.1 *Selective Hydrogenation of Alkynes, Alkenes, and Polyunsaturated Hydrocarbons*

The selective hydrogenation of alkynes and polyunsaturated hydrocarbons (e.g., dienes and ene-yne) to alkene is an important class of hydrogenations employed by bulk and fine chemical industry. The stringent purity requirements for polymer and chemical-grade alkenes imply that the concentration of highly unsaturated compounds should be reduced down to 5–10 ppm. Alkynes' and polyunsaturated hydrocarbons' impurities are effective poisons to alkene polymerization catalysts [42]. The catalytic partial hydrogenation of alkynes and polyunsaturated hydrocarbons to the corresponding mono-olefin is the most widely applied method to purify olefin streams. The semihydrogenation of C≡C-bonds to alkenes is one of the most useful hydrogenations for the production of vitamins and their intermediates, such as vitamins A and E.

The semihydrogenation of alkynes' reaction pathway is characterized by the sequential addition of hydrogen to the adsorbed alkyne or alkene, according to the Horiuti–Polanyi mechanism [43] (Figure 1.10). Theoretical calculations suggested that the adsorption strength of alkene is weaker than that of the alkyne (referred to as the thermodynamic factor) and is the primary factor for the identification of selective catalysts for alkyne semihydrogenation [44,45]. Briefly, the

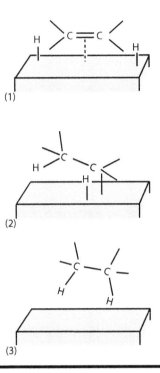

Figure 1.10   Steps in the hydrogenation of a C=C double bond at a catalyst surface—for example, Ni or Pt. (1) The reactants are adsorbed on the catalyst surface and H₂ dissociates. (2) An H atom bonds to one C atom. The other C atom is still attached to the surface. (3) A second C atom bonds to an H atom. The molecule leaves the surface.

mechanism starts by (1) binding of the unsaturated bond, and hydrogen dissociation into atomic hydrogen onto the catalyst followed by (2) addition of one atom of hydrogen, which is a reversible step, and finally (3) addition of the second atom—effectively irreversible under hydrogenating conditions. In the second step, the metallointermediate formed is a saturated compound that can rotate and then break down, again detaching the alkene from the catalyst. Consequently, contact with a hydrogenation catalyst necessarily causes *cis-trans*-isomerization, because the isomerization is thermodynamically favorable. This is a problem in partial hydrogenation, while in complete hydrogenation the produced *trans*-alkene is eventually hydrogenated.

1.2.1.1 Partial Hydrogenation of Alkyne Bonds

The partial hydrogenation of alkynes' carbon–carbon triple bonds to alkenes is an extremely useful hydrogenation for all quadrants of the chemical industry; however, careful choice of catalyst and reaction conditions is required to obtain high selectivity. In general, hydrogenation of acetylenes with a metal catalyst results in the formation of the fully saturated alkane product because the second hydrogenation (alkene to alkane) is generally faster than the first (alkyne to alkene). However, as long as some of the starting alkyne remains in the reaction mixture, selectivity can be high since the alkynes bind more strongly to the metal surface.

Catalyst modification can enhance selectivity. The main strategy has been to stop the second hydrogenation from occurring using suitable catalyst poisons, which modify the activity of the metal catalyst. The most notorious example is the catalyst developed by Lindlar [8,9,46], which is used world-wide. The catalyst's main components are palladium metal supported on calcium carbonate, which is doped with lead acetate solution during manufacture. This catalyst can then be used directly in the hydrogenation or modified further by an organic compound such as an amine.

One of the earliest uses of the catalyst developed by Lindlar was the partial-hydrogenation of a vitamin A key intermediate (Figure 1.11) to give tetraene. While this could be achieved with poisoned palladium on charcoal or palladium on calcium carbonate [47], selectivities were significantly higher with the Lindlar catalyst.

The Lindlar catalyst is now used in a multitude of reactions with particular importance in the synthesis of vitamins A and E, and also intermediates for the fragrance industry. An important starting material in DSM Nutritional Products' production of such compounds is methylbutenol, which is produced for the partial hydrogenation of the corresponding alkyne in a batch-wise process; selectivity is very high

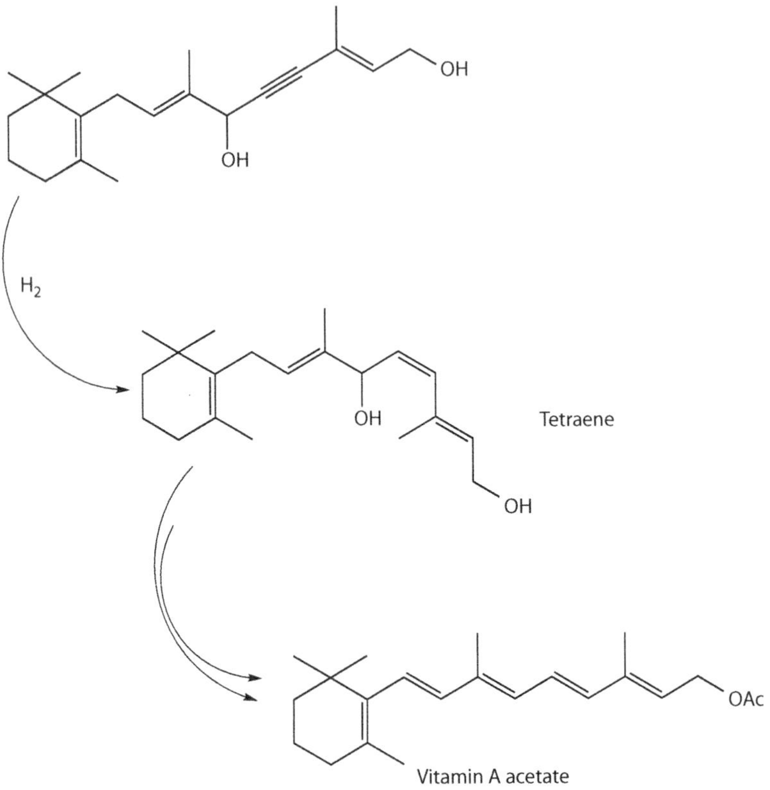

Figure 1.11   Partial hydrogenation of a vitamin A intermediate.

(>98%) and the catalyst can be recycled multiple times. From methylbutenol, the chain is extended in a sequential manner to obtain dehydroisophytol. This is then reduced in another semihydrogenation to give isophytol [48]. Isophytol can then be coupled with trismethylhydroquinone, to form α-tocopherol. As with previous reactions, the hydrogenation is carried out in a batch-wise process at 2–5 bar hydrogen pressure.

Two compounds of interest to the fragrance industry are linalool and linalyl acetate (Figure 1.12). Both have pleasant floral and spicy odors and are found in a wide range of natural flowers and spice plants. Their main uses are as perfume components in soaps, shampoos, and lotions. Their synthesis

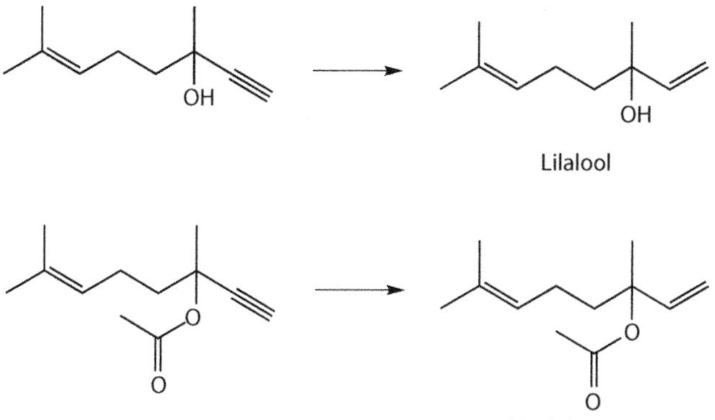

Figure 1.12   Synthesis of lilalool and linalyl acetate by partial hydrogenation of alkynes.

requires a partial hydrogenation of an alkyne precursor (see Figure 1.12), which is done with the Lindlar catalyst.

Another fragrance compound, dimethyloctenol, also has a selective semihydrogenation of a C≡C triple bond in its production scheme. The reaction is carried out with Lindlar catalysts. Because of the toxicity and health considerations posed by lead, scientists are trying to develop a new generation of Lindlar-type catalysts that solely contain palladium. The selectivity is controlled by careful preparation of palladium nanoparticles with narrow size distribution. These catalysts allow the hydrogenation of C≡C bonds with low metal loadings and, in several cases, with a high selectivity. An example of such lead-free reaction is the production of Z-butene-1,4-diol, which is an intermediate in the synthesis of vitamin B6, and is synthesized by the selective hydrogenation of butyne-diol in the presence of Pd supported on $CaCO_3$, carbon, or Al_2O_3 [49–51]. The use of supercritical fluids (e.g., sc-CO_2) has allowed such hydrogenations to be carried out in a continuous manner with a Pb-free system [52].

To conclude this section, I would like to emphasize a reaction scheme that can replace the current industrial process to produce resveratrol. Resveratrol is a polyphenol that can be isolated from natural sources such as red grapes or giant knotweed. It has attracted significant attention due to its possible health benefits in areas such as anticancer, antiaging, anti-inflammatory, and cardiovascular protection. The synthesis of resveratrol is carried out industrially using the Mizoroki–Heck coupling of 3,5-diacetoxy-styrene and 4-acetoxybromobenzene as a key step [53,54]. Alternatively, we can use partial hydrogenation of tolan derivatives, formed by Sonogashira coupling, to synthesize electron-rich stilbene derivatives (e.g., combretastatin) that can be subsequently converted to resveratrol (Figure 1.13). Preliminary results showed that Lindlar catalysts can be employed but the current concerns about lead make

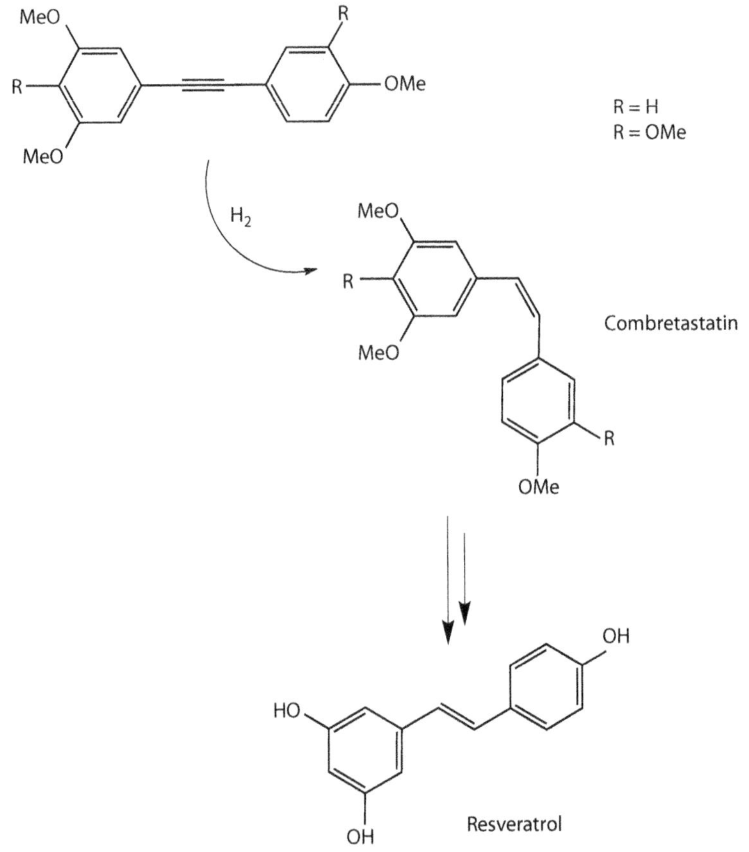

Figure 1.13 Alternative reaction scheme to the industrial process for the production of resveratrol.

the process less desirable [55]. Thus, research on alternative catalysts is in high demand.

1.2.1.2 Hydrogenation of Olefin Bonds

Carbon double bond hydrogenation is probably the most common hydrogenation in industry. A wide variety of catalysts are available from commercial suppliers and this transformation is considered a robust and atom economical reaction. While normal carbon double bond hydrogenations are carried out with palladium or platinum catalysts, stereoselective hydrogenation of such bonds employs homogenous catalysts composed of ruthenium and rhodium [56–63]. Heterogenization of such catalysts or heterogeneous catalysts all together is highly desirable for the referenced asymmetric reactions because it would simplify removal and recovery of the catalysts.

Due to its biological and antioxidant properties, (all-*rac*)-α-tocopherol is economically the most important member of the group of vitamin E compounds. This fat-soluble vitamin is produced on a scale of >30,000 tons per year for applications in human and animal nutrition. One of the key building blocks for the chemical production of synthetic vitamin E is trimethylhydroquinone, which is converted into (all-*rac*)-α-tocopherol by condensation with (all-*rac*)-isophytol [48,64,65]. There are two possible schemes for the production of trimethylhydroquinone, both involving Pd/C. The first is the hydrogenation of trimethylbenzoquinone and the other is from 2,6-dimethylbenzoquinone, which is first hydrogenated to 2,6-dimethylhydroquinone and methylated later by aminomethylation followed by hydrogenolysis (see Figure 1.14).

Another process that uses C=C hydrogenation for its synthesis is vitamin K_1. The core unit of all K-vitamins is menadione [66]. The standard synthesis of vitamins K_1 and K_2 is the coupling of the aromatic unit with the (poly)prenyl side chain. The direct coupling of the side chain to menadione is not possible; therefore a key step is the hydrogenation of menadione to menadiol

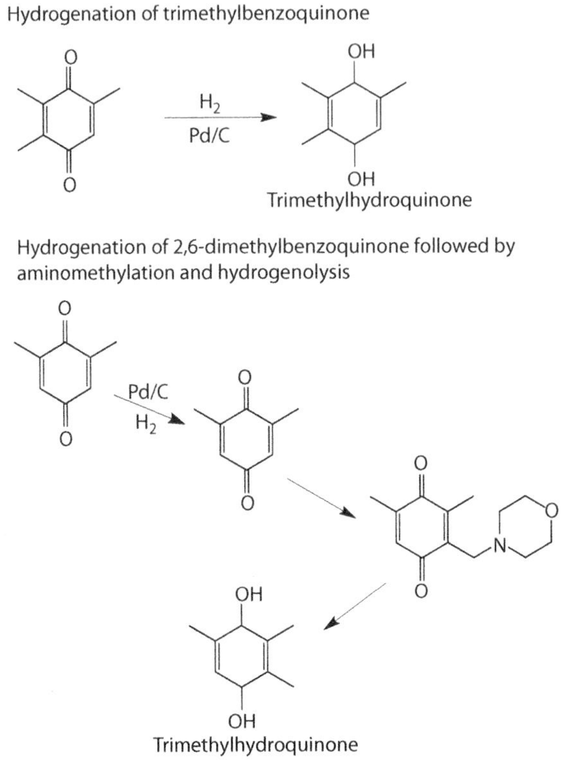

Hydrogenation of trimethylbenzoquinone

Trimethylhydroquinone

Hydrogenation of 2,6-dimethylbenzoquinone followed by aminomethylation and hydrogenolysis

Trimethylhydroquinone

Figure 1.14   Industrial schemes for the production of trimethylhydroquinone.

(Figure 1.15) so that alkylation can take place. The hydrogenation is carried out batch-wise using a palladium on a carbon catalyst.

As mentioned before, most of the asymmetric hydrogenation reactions are carried out by homogeneous catalysts containing rhodium or ruthenium. However, the stereoselective hydrogenation of a trisubstituted olefinic C=C double bond

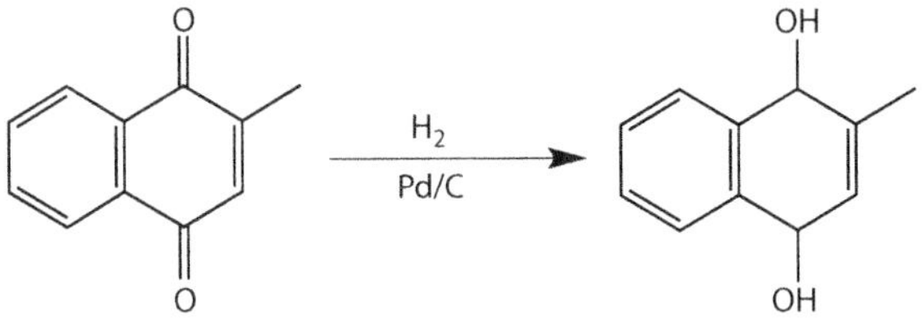

Figure 1.15   Hydrogenation of menadione to menadiol.

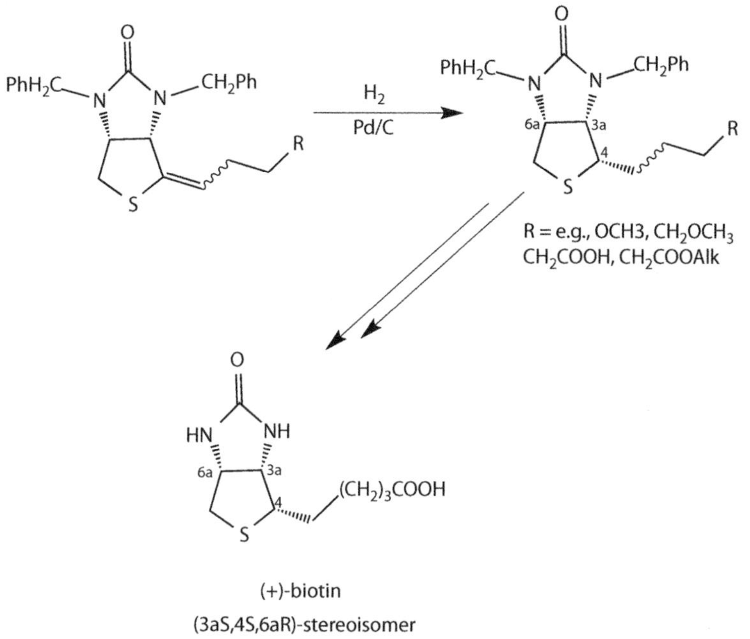

Figure 1.16   Hydrogenation step in vitamin (+)-biotin production.

vitamin (+)-biotin, produced on a scale of about 100 tons per year, can be catalyzed by Pd/C. The stereocenter C-4 of the thiophane ring can be introduced by catalytic hydrogenation of the exocyclic olefin with undefined double-bond stereochemistry on Pd/C or other heterogeneous catalysts, yielding the desired all-*cis* relative configuration at centers C-4, C-3a, and C-6a [67] (Figure 1.16).

1.2.2 Selective Hydrogenation of C=O and C≡N

The chemoselective hydrogenation of unsaturated aldehydes and ketones to unsaturated alcohols is an important class of reactions to produce a variety of fine chemicals [68–70] used as intermediates in the production of fragrances and flavors. Benzyl alcohol is a key intermediate in many chemical reactions, such as resveratrol production. Benzyl alcohol is conventionally produced from the hydrogenation of acetophenone

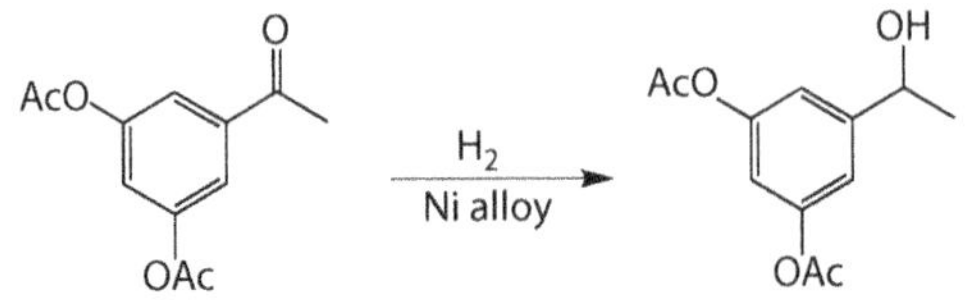

Figure 1.17 Hydrogenation of acetophenone derivative to benzyl alcohol derivative.

derivative using a transition metal catalyst [71] (Figure 1.17). A number of catalysts can successfully perform this transformation, but the suppression of by-products is essential. Palladium or platinum on carbon produced a high level of by-products, which was too high to be applied on a large scale. On the other hand, nickel-alloy catalysts in mild conditions yielded the required alcohol in very high yields (>95%) [71].

Catalytic hydrogenation of sugar is an atom-efficient transformation leading to the formation of high-value sugar alcohols (e.g., xylitol, sorbitol, arabitol). Xylitol and arabitol are safe sweeteners that are beneficial for dental health [72]. Driven by increasingly health- and weight-conscious consumers, xylitol and arabitol demand is expected to grow in sugar-free and low-calorie food products. The global market for xylitol is expected to reach 242,000 metric tons valued at just above US$1 billion by 2020. Currently, xylitol is manufactured via the chemical hydrogenation of xylose with a nickel catalyst at elevated temperature and pressure [73]. The process is laborious and cost and energy intensive. Alternative strategies based on noble metal catalysts (Pt, Ru) are not yet cost effective, but can be used with cellulose and hemicellulose. However, the process is considerably less efficient with hemicellulose [74]. Despite the recent developments in photocatalysis, to our knowledge there is no report of photoreduction of sugars into sugar alcohols.

L-Ascorbic acid (vitamin C) is produced from D-sorbitol, which is produced from the hydrogenation of D-glucose. The process can be carried out via microbiologic fermentation or electrochemical methods, but the preferred reaction scheme is hydrogenation using a nickel alloy catalyst branded under the name of Centoprime® (Figure 1.18) [75].

Furfural is a chemical building block for the production of transportation fuels as well as for a variety of useful acids, aldehydes, alcohols, and amines [76]. Furfural is produced from the dehydration of biomass sugars in the presence of an acid catalyst [77], which can be improved by the presence of ionic liquids [78]. Hydrogenation of furfural with hydrogen gas on metal catalysts results in the formation of furfuryl alcohol, which is primarily used as an ingredient in the manufacture of chemical products such as foundry resins, adhesives, and wetting agents [79].

The catalytic active center is composed of metals able to dissociate hydrogen, thus making hydrogenation possible. The choice of catalyst support is based on its ability to disperse and stabilize metal particles enhancing the active surface area. Compared to other aldehydes, in addition to the carbonyl group, furfural contains an aromatic furanyl ring that can be also hydrogenated. While carbonyl hydrogenation is usually preferred due to the high stability of the aromatic ring, metal catalysts that have strong interactions with the unsaturated C=C bonds can still saturate the ring. Thus, the selectivity toward aromatic alcohols is strongly dependent on the metal catalyst used. Furthermore, the geometric and electronic properties of different metals can affect both hydrogenation activity and selectivity by influencing the type of adsorption.

Copper has been most intensively investigated as a catalyst for furfural hydrogenation [80,81] (Figures 1.19 through 1.21). Silver

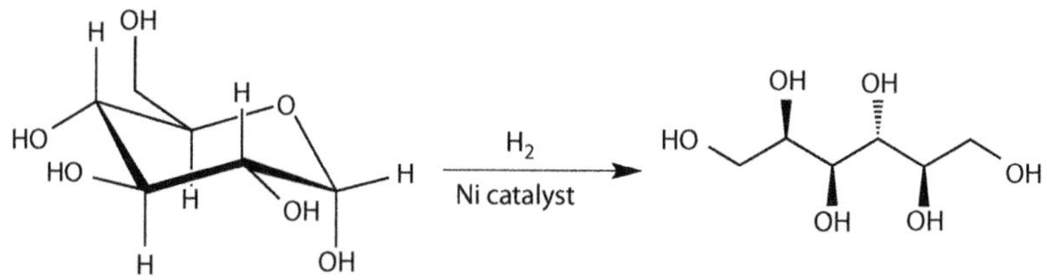

Figure 1.18 Hydrogenation of D-glucose to D-sorbitol.

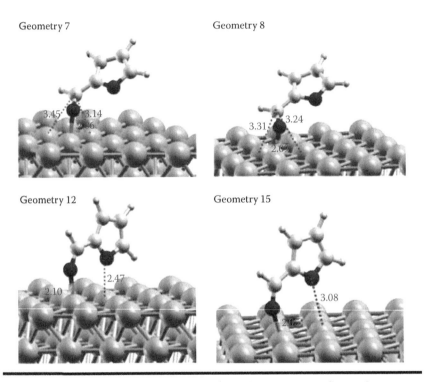

Figure 1.19 Optimized geometries for adsorption configuration on Cu(1 1 1) with higher adsorption energies. The distances shown in the figure are in angstroms. (Reproduced from S. Sitthisa, T. Sooknoi, Y. Ma, P. B. Balbuena, D. E. Resasco, *J. Catal.* 277 (2011) 1–13. With permission.)

has been used in a few studies, but there is no report on gold catalysts [82]. The group IB metals are significantly less active than other metals. However, they exhibit a remarkable selectivity toward hydrogenation of the carbonyl group, leaving the C=C double bonds in the furanyl ring unreacted. In this sense, Cu has been found to be the most selective among all tested metal catalysts. Selectivities above 98% to furfuryl alcohol have been achieved over monometallic Cu/SiO_2 catalysts [83].

Also, monometallic Ag catalysts have been found able to hydrogenate the C=O group of furfural with relatively good selectivity, but not as high as that of Cu. For example, 80% selectivity was obtained over Ag/SiO_2. Group VIII metals have

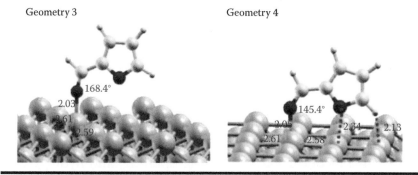

Figure 1.20 **Optimized geometries for adsorption configuration on Cu(1 1 0) with higher adsorption energies. The distances shown in the figure are in angstroms. (Reproduced from S. Sitthisa, T. Sooknoi, Y. Ma, P. B. Balbuena, D. E. Resasco,** *J. Catal.* **277 (2011) 1–13. With permission.)**

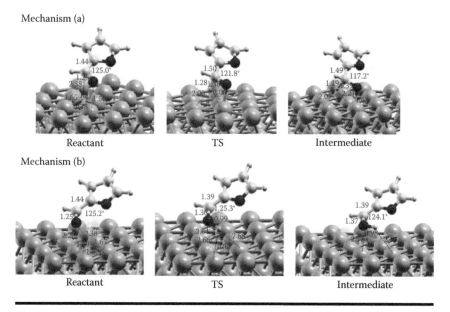

Figure 1.21 **Optimized geometries of the adsorbed furfural molecules, transition states, and intermediates Cu(1 1 1) via mechanisms (a) and (b). The distances shown in the figure are in angstroms. (Reproduced from S. Sitthisa, T. Sooknoi, Y. Ma, P. B. Balbuena, D. E. Resasco,** *J. Catal.* **277 (2011) 1–13. With permission.)**

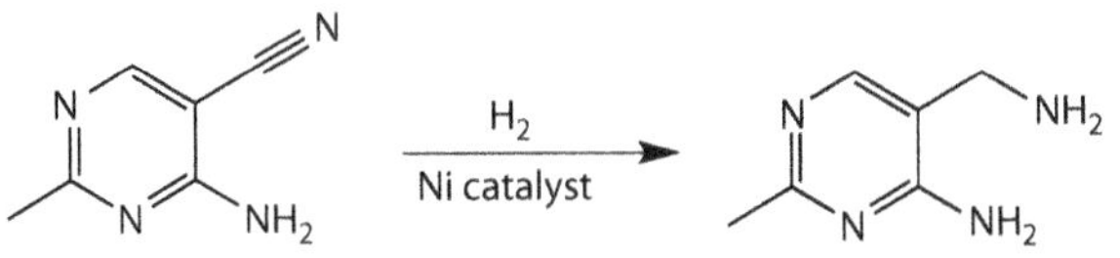

Figure 1.22   Hydrogenation of pyrimidino nitrile to Grewe diamine.

also been used in furfural hydrogenation, but they often have lower selectivities than group IB metals due to unwanted processes such as decarbonylation, ring opening, and hydrogenation of the furanyl ring. As a result, Cu and Ag are the choice of catalyst. The differences in activity and selectivity between group VIII and IB metals are related to substrate adsorption, which can be explained according to the aforementioned Nørskov model on the basis of d-orbital filling. In-group IB metals' d-orbitals are filled, reducing the bond strength [84].

Vitamin B1 (thiamine chloride) contains two heterocyclic rings linked by a methylene unit. In all industrial syntheses, the key intermediate is Grewe diamine, which is prepared by hydrogenation of a pyrimidino nitrile using a nickel-alloy catalyst (Figure 1.22). It was reported for the first time in 1944 by Huber [85], who found that the use of palladium- and platinum-supported catalysts gave significant amounts of secondary amines as by-products. The use of nickel catalysts also gave significant quantities of secondary amines, but these could be reduced to less than 5% by the addition of ammonia to the reaction mixture.

1.3 Concluding Remarks

As mentioned before, this chapter is not aimed to be a thorough review of the field but rather to highlight the importance of heterogeneous catalysis for fine chemical production and to set the motif for the subsequent chapters. The reactions highlighted are by no means the only ones, but they provide a flavor of what it is being done and what more can be done in research and development of catalytic systems. The prime emphasis was on the field of vitamins and fragrances since they represent industries forecasted for uninterrupted growth over the next decades. However, novel and improved catalysts are needed in all quadrants. The demand for sustainability and greener processes also means that novel routes will be created that take advantage of biorenewable raw materials, which also demand more and better catalytic systems.

Economical and environmental factors are on the same side when it comes to what the catalysts of the future should be. They should be based on abundant and nontoxic metals, primarily copper, iron, and so on. Their usage was made possible due to advances in nanoscience. In this length scale, metals display not only improved catalytic properties, but also unexpected and unusual reactivity and selectivity. Thus, for any person involved or planning to be involved in this area in the future, the sky is very blue (or green if you prefer), with many and unlimited opportunities. Also, as has been the case so far, catalyst development is a multidisciplinary approach, which makes it rather an unique and interesting field of both academic and industrial research.

References

1. S. Nishimura, *Handbook of heterogeneous catalytic hydrogenation for organic synthesis*, Wiley-Interscience (2001).
2. E. Klabunovskii, G. V. Smith, Á. Zsigmond, *Heterogeneous enantioselective hydrogenation: 31 (catalysis by metal complexes)*, Springer Netherlands (2007).
3. K. Wilson, A. F. Lee, *Heterogeneous catalysts for clean technology: Spectroscopy, design, and monitoring*, Wiley-VCH (2013).
4. N. Giusnet, *Heterogeneous catalysis and fine chemicals: Proceedings (Studies in Surface Science and Catalysis)*, Elsevier Science Ltd. (1988).
5. M. Yan, *Development of new catalytic performance of nanoporous metals for organic reactions*, Springer Japan (2014).

6. G. V. Smith, F. Notheisz, *Heterogeneous catalysis in organic chemistry*, Academic Press (1999).

7. P. Sabatier, *Comptes Rendus Hebdomadaires des Seances de l'Academie des Science* 128 (1899) 1173.

8. H. Lindlar, Ein neuer Katalysator fur selektive Hydrierungen, *Helv. Chim. Acta* 35 (1952) 446–450.

9. H. Lindlar, R. Dubuis, Palladium catalyst for partial reduction of acetylenes, *Org. Synth.* 45 (1966) 89–92.

10. W. S. Knowles, Asymmetric hydrogenations (Nobel lecture), *Angew. Chem. Int. Ed.* 41 (2002) 1998.

11. R. Noyori, Asymmetric catalysis: Science and opportunities (Nobel lecture), *Angew. Chem. Int. Ed.* 41 (2002) 2008.

12. R. Hoffmann, Dobereiner's lighter, *Am. Scientist* 86 (1998) 326.

13. M. Appl, Ammonia, In *Ullmann's encyclopedia of industrial chemistry*, Wiley-VCH (2006).

14. F. Bozso, G. Ertl, M. Grunze, M. Weiss, Interaction of nitrogen with iron surfaces. I. Fe(100) and Fe(111), *J. Catal.* 49 (1977) 18–41.

15. R. Imbihl, R. J. Behm, G. Ertl, W. Moritz, The structure of atomic nitrogen on Fe(100), *Surf. Sci.* 123 (1982) 129–140.

16. G. Ertl, S. B. Lee, M. Weiss, Kinetics of nitrogen adsorption on Fe(111), *Surf. Sci.* 114 (1982) 515–526.

17. G. Ertl, Primary steps in catalytic synthesis of ammonia. *J. Vacuum Sci.* 1 (1983) 1247–1253.

18. G. Ertl, Surface science and catalysis—Studies on the mechanism of ammonia synthesis: The P. H. Emmett Award Address, *Catal. Rev. Sci. Eng.* 21 (1980) 201–223.

19. H. Schulz, Short history and present trends of Fischer-Tropsch synthesis, *Appl. Catal. A* 186 (1999) 3–12.

20. M. Raney, Method of preparing catalytic material, US Patent 1563587, issued 1925-12-01.

21. M. Raney, Method of producing finely divided nickel, US Patent 1628190, issued 1927-05-10.

22. P. N. Rylander, Hydrogenation and dehydrogenation, In *Ullmann's encyclopedia of industrial chemistry*, Wiley-VCH (2005).

23. M. Herskowitz, Modelling of a trickle-bed reactor—The hydrogenation of xylose to xylitol, *Chem. Eng. Sci.* 40 (1985) 1309–1311.

24. I. P. Freeman, Margarines and shortenings, In *Ullmann's encyclopedia of industrial chemistry*, Wiley-VCH (2005).

25. M. Crespo-Quesada, A. Yarulin, M. Jin, Y. Xia, L. Kiwi-Minsker, Structure sensitivity of alkynol hydrogenation on shape- and size-controlled palladium nanocrystals: Which sites are most active and selective? *J. Am. Chem. Soc.* 133 (2011) 12787–12794.
26. E. Schmidt, A. Vargas, T. Mallat, A. Baiker, Shape-selective enantioselective hydrogenation on Pt nanoparticles, *J. Am. Chem. Soc.* 131 (2009) 12358–12367.
27. B. Hammer, J. K. Nørskov, Why gold is the noblest of all the metals, *Nature* 376 (1995) 238.
28. B. Hammer, J. K. Nørskov, Atomic chemisorption, *Adv. Catal.* 45 (2000) 71.
29. F. H. B. Lima, J. Zhang, M. H. Shao, K. Sasaki, M. B. Vukmirovic, E. A. Ticianelli, R. R. Adzic, Catalytic activity—d-band center correlation for the O_2 reduction reaction on Pt in alkaline solutions, *J. Phys. Chem. C* 111 (2007) 404.
30. U. Heiz, E. L. Bullock, Fundamental aspects of catalysis on supported metal clusters, *J. Mater. Chem.* 14 (2004) 564.
31. W. M. C. Sameera, F. Maseras, Transition metal catalysis by density functional theory and density functional theory molecular mechanics, *WIREs Comput. Mol. Sci.* 2 (2012) 375.
32. H. G. Manyar, B. Yang, H. Daly, H. Moor, S. McMonagle, Y. Tao, G. D. Yadav, A. Goguet, P. Hu, C. Hardacre, Selective hydrogenation of α,β-unsaturated aldehydes and ketones using novel manganese oxide and platinum supported on manganese oxide octahedral molecular sieves as catalysts, *Chem. Cat. Chem.* 5 (2013) 506–512.
33. H. G. Manyar, R. Morgan, K. Morgan, B. Yang, P. Hu, J. Szlachetko, J. Sá, C. Hardacre, High energy resolution fluorescence detection XANES—An in situ method to study the interaction of adsorbed molecules with metal catalysts in the liquid phase, *Catal. Sci. Technol.* 3 (2013) 1497.
34. M. Makosch, C. Kartusch, J. Sa, R. B. Duarte, J. A. van Bokhoven, K. Kvashnina, P. Glatzel, J. Szlachetko, D. L. A. Fernandes, E. Kleymenov, M. Nachtegaal, K. Hungerbühler, HERFD XAS/ATR-FTIR batch reactor cell, *Phys. Chem. Chem. Phys.* 14 (2012) 2164.
35. W. Yu, M. D. Porosoff, J. G. Chen, Review of Pt-based bimetallic catalysis: From model surfaces to supported catalysts, *Chem. Rev.* 112 (2012) 5780–5817.

36. S. T. Marshall, M. O'Brien, B. Oetter, A. Corpuz, R. M. Richards, D. K. Schwartz, J. W. Medlin, Controlled selectivity for palladium catalysts using self-assembled monolayers, *Nat. Mater.* 9 (2010) 853–858.

37. M. Makosch, W.-I. Lin, V. Bumbálek, J. Sá, J. W. Medlin, K. Hungerbühler, J. A. van Bokhoven, Organic thiol modified Pt/TiO$_2$ catalysts to control chemoselective hydrogenation of substituted nitroarenes, *ACS Catal.* 2 (2012) 2079–2081.

38. F. Nerozzi, Heterogeneous catalytic hydrogenation, *Platinum Metals Rev.* 56 (2012) 236–242.

39. W. Bonrath T. Netscher, Catalytic processes in vitamins synthesis and production, *Appl. Catal. A* 280 (2005) 55–73.

40. R. A. Sheldon, H. van Bekkum, *Fine chemicals through heterogeneous catalysis*, Wiley-VCH (2001).

41. R. A. Sheldon, Atom efficiency and catalysis in organic synthesis, *Pure Appl. Chem.* 72 (2000) 1233–1246.

42. A. Borodziński, G. C. Bond, Selective hydrogenation of ethyne in ethene-rich streams on palladium catalysts. Part 1. Effect of changes to the catalyst during reaction, *Catal. Rev. Sci. Eng.* 48 (2006) 91.

43. I. Horiuti, M. Polanyi, Exchange reactions of hydrogen on metallic catalysts, *Trans. Faraday Soc.* 30 (1934) 1164–1172.

44. F. Studt, F. Abild-Pedersen, T. Bligaard, R. Z. Sorensen, C. H. Chistensen, J. K. Nørskov, On the role of surface modifications of palladium catalysts in the selective hydrogenation of acetylene, *Angew. Chem. Int. Ed.* 47 (2008) 9299–9302.

45. F. Studt, F. Abild-Pedersen, T. Bligaard, R. Z. Sorensen, C. H. Chistensen, J. K. Nørskov, Identification of non-precious metal alloy catalysts for selective hydrogenation of acetylene, *Science* 320 (2008) 1320–1322.

46. H. Lindlar, Hydrogenation of acetylenic bond utilizing a palladium-lead catalyst (1954) US2681938.

47. O. Isler, W. Huber, A. Ronco, M. Kofler, Synthese des vitamin A, *Helv. Chim. Acta* 30 (1947) 1911–1927.

48. W. Bonrath, M. Eggersdorfer, T. Netscher, Catalysis in the industrial preparation of vitamins and nutraceuticals, *Catal. Today* 121 (2007) 45–57.

49. H. Pauling, B. J. Weimann, Vitamin B6, In *Ullmann's encyclopedia of industrial chemistry*, Willey-VCH, (1996) 530–540.

50. T. Fukuda, T. Kusama, Partial hydrogenation of 1,4-butynediol, *Bull. Chem. Soc. Jpn.* 31 (1958) 339–342.

51. J. M. Winterbottom, H. Marwan, E. H. Stitt, R. Natividad, The palladium catalysed hydrogenation of 2-butyne-14-diol in a monolith bubble column reactor, *Catal. Today* 79–80 (2003) 391–399.
52. R. Tschan, M. M. Schubert, A. Baiker, W. Bonrath, H. Lansink-Rotgerink, Continuous semihydrogenation of a propargylic alcohol over amorphous Pd_{18}/Si_{19} in dense carbon dioxide: Effect of modifiers, *Catal. Lett.* 75 (2001) 31–36.
53. B. Wüstenberg, R. T. Stemmler, U. Letinois, W. Bonrath, M. Hugentobler, T. Netscher, Large-scale production of bioactive ingredients as supplements for healthy human and animal nutrition, *Chimia* 65 (2011) 420–428.
54. R. Haerter, U. Lemke, A. Radspieler, Process for the preparation of stilbene derivatives (2005) WO2005023740.
55. U. Letinois, W. Bonrath, Preparation of substituted electron rich diphenylacetylens and stilbene derivatives (2011) WO201109888.
56. T. Netscher, M. Scalone, R. Schmid, Enantioselective hydrogenation: Towards a large-scale total synthesis of (R,R,R)-α-tocopherol, eds. H.-U. Blaser, E. Schmidt, *Asymmetric catalysis on industrial scale*, Wiley-VCH (2004) 71–89.
57. S. Bell, B. Wüstenberg, S. Kaiser, F. Menges, T. Netscher, A. Pfaltz, Asymmetric hydrogenation of unfunctionalized, purely alkyl-substituted olefins, *Science* 311 (2006) 642–644.
58. R. Imwinkelried, Catalytic asymmetric hydrogenation in the manufacture of d-biotin and dextrometorphan, *Chimia* 51 (1997) 300–302.
59. W. Bonrath, R. Karge, T. Netscher, F. Roessler, F. Spindler, Biotin—The chiral challenge, *Chimia* 63 (2009) 265–269.
60. K. Matsuki, H. Inoue, M. Takeda, Highly enantioselective reduction of meso-1,2-dicarboxylic anhydrides, *Tetrahedron Lett.* 34 (1993) 1167–1170.
61. K. Kaiser, B. de Potzolli, Pantothenic acid, In *Ullmann's encyclopedia of industrial chemistry*, Willey-VCH (1996) 559–566.
62. R. Schmid, Homogeneous catalysis with metal complexes in a pharmaceuticals' and vitamins' company: Why, what for, and where to go? *Chimia* 50 (1996) 110–113.
63. M. Schürch, N. Künzle, T. Mallat, A. Baiker, Enantioselective hydrogenation of ketopantolactone: Effect of stereospecific product crystallization during reaction, *J. Catal.* 176 (1998) 569–571.

64. K.-U. Baldenius, L. von dem Bussche-Hünnefeld, E. Hilgemann, P. Hoppe, R. Stürmer, Vitamin E (tocopherols, tocotrienols), In *Ullmann's encyclopedia of industrial chemistry*, Willey-VCH (1996) 478–488.

65. T. Netscher, Synthesis of vitamin E, *Vitamins and Hormones* 76 (2007) 155–202.

66. A. Rüttimann, Recent advances in the synthesis of K-vitamins, *Chimia* 40 (1986) 290–306.

67. M. W. Goldberg, L. H. Sternbach, Synthesis of biotin, (1949) US 2489232 and US 2489235.

68. J. Jenck, J. E. Germain, High-pressure competitive hydrogenation of aldehydes, ketones, and olefins on copper chromite catalyst, *J. Catal.* 65 (1980) 141.

69. W. H. Carothers R. Adams, Platinum oxide as a catalyst in the reduction of organic compounds. VII. A study of the effects of numerous substances on the platinum catalysis of the reduction of benzaldehyde, *J. Am. Chem. Soc.* 47 (1925) 1047.

70. P. Gallezot, D. Richard, Selective hydrogenation of α,β-unsaturated aldehydes, *Catal. Rev. Sci. Eng.* 40 (1980) 81.

71. W. Bonrath, T. Müller, L. Kiwi-Minsker, A. Renken, I. Iourov, Hydrogenation process of alkynols to alkenols in the presence of structured catalysts based on sintered metal fibers (2011) WO2011092280.

72. L. M. Steinberg, F. Odusola, I. D. Mandel, Remineralizing potential, antiplaque and antigingivitis effects of xylitol and sorbitol sweetened chewing gum, *Clin. Prev. Dent.* 14 (1992) 31–34.

73. J. Wisniak, M. Hershkowitz, R. Leibowitz, S. Stein, Hydrogenation of xylose to xylitol, *Ind. Eng. Chem. Prod. Res. Dev.* 13 (1974) 75.

74. H. Kobayashi, Y. Yamakoshi, Y. Hosaka, Mi. Yabushita, A. Fukuoka, Production of sugar alcohols from real biomass by supported platinum catalyst, *Catal. Today* 226 (2014) 204–209.

75. https://catalysts.evonik.com/sites/dc/Downloadcenter/Evonik/Product/Catalysts/Publications/ArticleCatalysts_en.pdf.

76. F. W. Lichtenthaler, Unsaturated O- and N-heterocycles from carbohydrate feedstocks, *Acc. Chem. Res.* 35 (2002) 728.

77. C. D. Hurd, L. L. Isenhour, Pentoses reactions. I. Furfural formation, *J. Am. Chem. Soc.* 54 (1932) 317.

78. H. Zhao, J. E. Holladay, H, Brown, Z. C. Zhang, Metal chlorides in ionic liquid solvents convert sugars to 5-hydroxymethylfurfural, *Science* 316 (2007) 1597–1600.

79. K. J. Zeitsch, *The chemistry and technology of furfural and its many by-products,* Elsevier Science, Amsterdam (2000).
80. G. Seo, H. Chon, Hydrogenation of furfural over copper-containing catalysts, *J. Catal.* 67 (1981) 424.
81. R. Rao, A. Dandekar, R. T. K. Baker, M. A. Vannice, Properties of copper chromite catalysts in hydrogenation reactions, *J. Catal.* 171 (1997) 406–419.
82. P. Claus, Selective hydrogenation of α,β-unsaturated aldehydes and other C=O and C=C bonds containing compounds, *Top. Catal.* 5 (1998) 51.
83. S. Sitthisa, T. Sooknoi, Y. Ma, P. B. Balbuena, D. E. Resasco, Kinetics and mechanism of hydrogenation of furfural on Cu/SiO$_2$ catalysts, *J. Catal.* 277 (2011) 1–13.
84. M. K. Bradley, J. Robinson, D. P. Woodruff, The structure and bonding of furan on Pd(111), *Surf. Sci.* 604 (2010) 920–925.
85. W. Huber, Hydrogenation of basic nitriles with Raney nickel, *J. Am. Chem. Soc.* 66 (1944) 876–879.

Chapter 2

Hydrogenation by Nickel Catalysts

Anna Śrębowata and Jacinto Sá

Contents

This chapter deals with hydrogenation mediated by nickel metal presence as supported and unsupported nanoparticles. In the chapter, we highlight synthetic procedures for the preparation of metallic nickel nanoparticles, and their applications to the catalytic reactions important in technological and environmental points of view. The main target is reactions carried out for fine and pharmaceutical chemical production.

2.1 Introduction

Nickel has been found in metallic artifacts from more than 2000 years ago. It was identified and isolated as an element for the first time by Axel Cronstedt, a Swedish chemist, in 1751. In the nineteenth century, it came to prominence in plating and in alloys such as "nickel silver" (German silver) in which nickel is alloyed with copper and zinc. Its name comes from the Saxon term "kupfernickel" or "devil's copper," as the fifteenth century miners thought the ore looked red-brown like copper but it was too difficult to mine and was poisoning them. Later, arsenic was found to be in the ore. Coins in the United States first used nickel alloyed with copper in 1857. The "nickel" was not pure nickel, but in 1881, pure nickel was used for coins in Switzerland. Stainless steels were discovered early in the twentieth century and nickel was found to have a very beneficial role in many of the common grades, a situation that continues to this day. Alloys based on nickel were found to have excellent corrosion resistance and high temperature resistance, which made them suitable for chemical plants and also allowed the practical realization of the jet engine. As a result of these developments, nickel enjoyed a very strong growth of demand in the twentieth century and continues to do so.

Nickel-containing materials are used in many applications, such as food preparation equipment, mobile phones, medical equipment, transport, buildings, and power generation, because they provide the desired combination of mechanical properties, corrosion resistance, physical properties, durability, appearance, availability, ease of use, and economic performance throughout their life cycle. Nickel-containing alloys are used extensively in the chemical, pharmaceutical, and petrochemical industries for their corrosion resistance to both aqueous and gaseous conditions in high-temperature environments, their mechanical properties at all temperatures from cryogenic to the very high, and, occasionally, for their special physical properties. Nickel plating is used for its special properties, such as hardness or corrosion resistance. Nickel plays an important role in nature, too. Enzymes of some microorganisms and plants contain nickel as an active site, which makes the metal an essential nutrient for them.

This chapter deals with the application of nickel-containing materials as the catalysts in hydrogenation processes. Application of nickel in hydrogenation reactions is almost as old as this reaction itself. The first known application of nickel in hydrogenation dates back to 1897, when French chemist Paul Sabatier carried out the reaction of hydrogen with carbon dioxide at elevated temperatures (optimally 300°C–400°C), and pressures in the presence of a nickel catalyst to produce methane and water:

$$CO_2 + 4H_2 \rightarrow CH_4 + 2H_2O + \text{energy}$$

$$\Delta H = -165.0 \text{ kJ/mol}$$

This process was named the "Sabatier reaction" and its author received the 1912 Nobel Prize in chemistry. A few years later, Wilhelm Normann used dispersed nickel as a catalyst for hydrogenation of liquid oleic acid into solid stearic acid. And, in 1924, Murray Raney developed a nickel fine powder catalyst, which was named after him, that is still widely used in hydrogenation reactions such as conversion of nitriles to amines or the production of margarine. Since the beginning of the twentieth century, the popularity of the nickel catalyst has not waned.

The application of nickel catalysts plays a crucial role in industrial technology such as nylon production (hydrogenation of benzene toward cyclohexane) and ammonia synthesis (the principal form of nitrogen used in fertilizers). Production of oils and fats derived from natural sources, such as palm and vegetable oils, depends on nickel-catalyst technology, too. Nickel-based catalysts play a major role in the production of surfactants from petrochemical and oleochemical (natural oils) feedstock. On the other hand, nickel catalysts also play a complex and extensive role in underpinning the competitiveness of the oil-refining industry. Finally, nickel facilitates greater flexibility of choice of input materials by refiners.

However, productivity (performance) of technological processes depends on the efficiency of nickel-catalyzed steps like hydrodenitrogenation to reduce NO_x and hydrodesulfurization to reduce SO_x in petroleum refining, selective hydrogenation of the C=C double bond, and hydrogenation of benzene toward cyclohexane. Interest in use of nickel in processes related to environmental protection is steadily growing, too. Nickel catalysts are investigated in purification of air and water—for example, hydrotreatment of volatile organic compounds (VOCs). In view of the extensive application of nickel in hydrogenation processes, this chapter is focused on the heterogeneous catalytic reactions important from a practical point of view. However, there are significant developments in the homogeneous fields that readers can find elsewhere [1].

2.2 Hydrogenation of Unsaturated Functional Groups

The main aim of this chapter is to describe the properties of the nickel catalyst in chemoselective hydrogenation reactions,

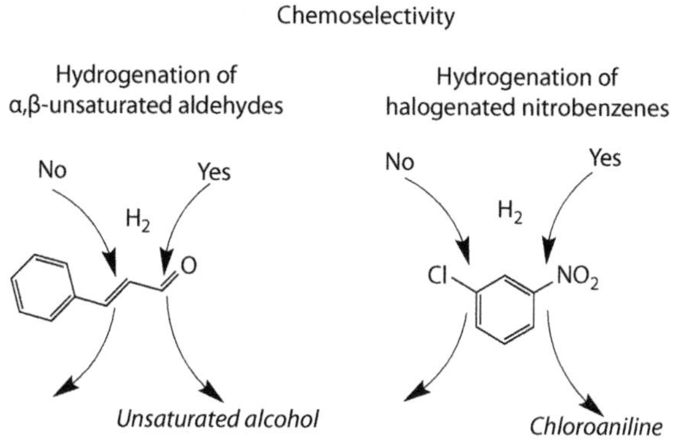

Chemoselectivity

| Hydrogenation of | Hydrogenation of |
| α,β-unsaturated aldehydes | halogenated nitrobenzenes |

Figure 2.1   Concept for chemoselective hydrogenation reactions as a method for the production of high-value chemicals.

where the nickel is capable of discriminating between different functional groups in the molecule and converting only one of them (i.e., chemoselective hydrogenation). The general idea of chemoselective hydrogenation is shown in Figure 2.1. The most important examples of this type of hydrogenation reactions are a hydrogenation carbonyl group in the presence of a C=C double bond and hydrogenation of a nitric group in the presence of C–Cl bonds. Chemoselective hydrogenation is an efficient way to obtain technologically significant products.

2.2.1 *Hydrogenation of α,β-Unsaturated Aldehydes*

Selective hydrogenation of α,β-unsaturated aldehydes to unsaturated alcohols is an essential step in the synthesis of chemical intermediates used in the production of pharmaceuticals, cosmetics, and foods [2]. The development of heterogeneous catalysts capable of selective conversion of α,β-unsaturated compounds would improve process efficiency. However, several challenges appear in the selective hydrogenation of α,β-unsaturated carbonyl compounds. The process of hydrogenation of α,β-unsaturated aldehydes can lead to the formation of unsaturated alcohols (desired products), saturated aldehydes, or saturated alcohols. The addition of hydrogen to the conjugated C=C bond is both thermodynamically and kineti-cally favored over the C=O bond, due to the lower C=C bond's average dissociation enthalpy (611 kJ/mol) than for C=O bond (737 kJ/mol) [3]. Nevertheless, α,β-unsaturated aldehydes and ketones can sometimes be selectively hydrogenated to the unsaturated alcohols in the presence of certain catalysts at suf-ficiently low temperature and pressure.

The steric hindrance associated with bulky substituent groups near the C=C bond is also thought to play an impor-tant role in determining selectivity, such as for ketones pres-ent in ring structures [4]. Steric hindrance near the C=C bond cannot alone explain why there is a lack of selectivity to the unsaturated alcohol for most α,β-unsaturated ketone hydro-genation over Pd, Pt, Rh, and Ru catalysts, since these metals show activity in selective hydrogenation of α,β-unsaturated aldehydes with similar sterically hindered C=C bonds. Among a plethora of α,β-unsaturated aldehydes, acrolein and cinna-maldehyde are the most popular and their hydrogenation is an important process in the technological point of view. Until now not a lot of attention has been paid to transition metals in chemoselective hydrogenation of α,β-unsaturated aldehydes.

2.2.1.1 Hydrogenation of Acrolein

Acrolein is one of the simplest α,β-unsaturated aldehydes, so it is often used as a model compound in the hydrogenation process. Luo et al. [5] performed theoretical calculations on acrolein hydrogenation via allyl alcohol, propanal, or enol to propanol on the Ni(111) surface, using the spin-polarized peri-odic density functional theory method and taking into account all possible intermediates and products as well as their cor-responding transition states. Figure 2.2 shows the most stable adsorption configurations along with selected bond parameters for the hydrogenation intermediates from acrolein to propa-nol. This most stable acrolein adsorption configuration has the C=O and C=C bonds over two neighboring face-centered

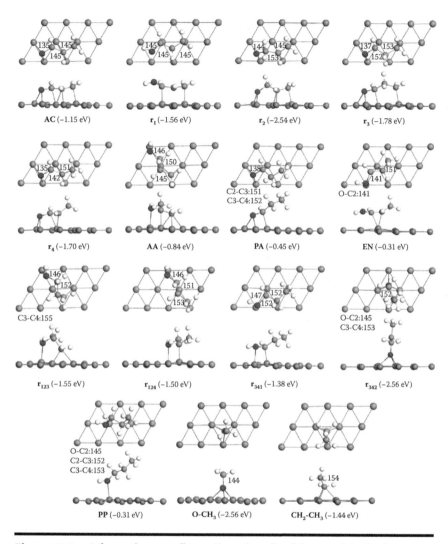

Figure 2.2   Adsorption configurations involved in acrolein hydrogenation into propanol on Ni(111) (bond lengths in picometers). (Reproduced from Q. Luo, T. Wang, M. Beller, H. Jiao, *J. Phys. Chem. C* 117 (2013) 12715–12724. With permission.)

cubic (fcc) sites, and the C=O and C=C bonds are elongated to 135 and 145 pm, compared to the gas-phase values (123 and 134 pm).

Luo et al. [5] showed that acrolein has an adsorption energy similar to molecular hydrogen on Ni(111). Therefore, the coadsorption of both molecules on the catalysts' surface is possible and the α,β-unsaturated aldehyde hydrogenation could be interpreted via the Langmuir–Hinshelwood mechanism on the surface. Additionally, theoretical calculations showed that the formation of propanal on Ni(111) is thermodynamically and kinetically favored, whereas the reaction path into allyl alcohol is more endothermic with a characteristically high energetic barrier.

Density functional theory (DFT) studies on selective acrolein hydrogenation on Ni(111) confirmed that the C=C double bonds can be more easily hydrogenated both kinetically and thermodynamically than C=O double bonds in acrolein hydrogenation to propanal and allyl alcohol. The results obtained by theoretical calculations are in agreement with results of experiments on nickel/support systems. On the other hand, Coq et al. [6] have shown good initial selectivity on nickel toward allyl alcohol.

2.2.1.2 Hydrogenation of Cinnamaldehyde

Cinnamaldehyde is one of the most popular α,β-unsaturated aldehydes; it is important in industry because it is the precious starting material for synthesis of fine chemicals. Its selective hydrogenation products are widely used (e.g., in the perfume industry and preparation of the pharmaceuticals used in the treatment of HIV). The final product of cinnamaldehyde hydrogenation is hydrocinnamyl alcohol, and this process can proceed via the formation of hydrocinnamaldehyde or cinnamyl alcohol. Because the first route is thermodynamically more favorable, the real challenge is to convert cinnamaldehyde toward cinnamyl alcohol. A potential reaction pathway for cinnamaldehyde hydrogenation is presented in Figure 2.3.

The product distribution strongly depends on the used catalyst. Noble metals such as platinum or gold are widely used in hydrogenation of cinnamaldehyde to cinnamyl alcohol [7,8]. On the other hand, the application of palladium or ruthenium

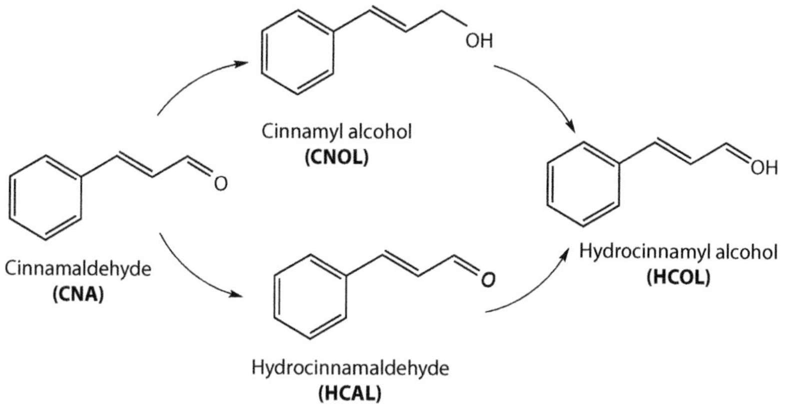

Figure 2.3 Reaction pathways for hydrogenation of cinnamaldehyde.

catalysts led to the formation of hydrocinnamaldehyde [9,10]. Until now not a lot of attention has been paid to monometallic nickel as the catalyst for the chemoselective hydrogenation of cinnamaldehyde [11–14]. Generally, application of nickel as the catalyst in hydrogenation of cinnamaldehyde both in gas and liquid phases leads to formation of hydrocinnamaldehyde, a compound that can be used, for example, in a range of fragrances. On the other hand, Gryglewicz et al. [15] showed that the activity and product distribution in cinnamaldehyde hydrogenation strongly depends on the nickel particles' size distribution and the kind of carbon support. Better performance in the liquid-phase hydrogenation of cinnamaldehyde to hydrocinnamaldehyde was measured when small nanoparticles were used. It was found that Ni supported on carbon nanofibers shows the highest catalytic activity in the hydrogenation reaction.

The effect of the nickel catalyst preparation method in the catalytic performance was investigated by Viswanathan et al. [11]. Ni/TiO$_2$ catalysts were prepared by four different methods, such as direct impregnation (IM), deposition precipitation (DP) with urea, and by chemical reduction using hydrazine hydrate (HH) and glucose (GL) as reducing agents exhibited good cinnamaldehyde conversion. The authors showed that crystallite size effects on activity are not expected in the size range (10–20 nm). Additionally, all four catalysts displayed significant but moderate selectivity toward cinnamyl alcohol. Catalysts prepared by HH and GL, with the smallest nickel particle size, displayed higher cinnamaldehyde conversion and selectivity to cinnamyl alcohol. Conversion of hydrocinnamaldehyde to hydrocinnamyl alcohol proceeds at a slower rate than that for cinnamyl alcohol to hydrocinnamyl alcohol.

The same authors proposed a probable mechanism of cinnamaldehyde adsorption on the nickel surface. Rather unexpected high selectivity to cinnamyl alcohol for the nickel catalyst was explained by the adsorption of cinnamaldehyde through the C=O bond, which is facilitated by the presence of small amounts of Ni^{2+}, along with Ni^0. The nature of metal-support interactions, which vary depending on the method of preparation (HH and GL), and the use of methanol as solvent could be the other contributing factors for the higher selectivity to cinnamyl alcohol.

The effect of nickel dispersion was discussed by Malobela et al. [16]. The turnover frequency (TOF) of the nickel catalysts and the selectivity to unsaturated alcohol increased in the following order: graphite < activated carbon < multiwalled carbon nanotubes. This phenomenon was strongly related with the metal particle size, which increases in reverse order: multiwalled carbon nanotubes (2.8 nm) < activated carbon (7.8 nm) < graphite (14.5 nm).

In the vapor phase hydrogenation of cinamaldehyde over 5%–15% $Ni/\gamma-Al_2O_3$ catalysts, the conversion increases apparently with the Ni surface area and the contact time, and the product selectivity depends on the level of conversion. Additionally, the stability of TOF was showed. This observation suggests that cinnamaldehyde hydrogenation over 10% $Ni/\gamma-Al_2O_3$ catalysts is a structure-insensitive reaction [17].

Fan et al. published important insights on cinnamaldehyde hydrogenation [18]. They determined the optimal hydrogenation of cinnamaldehyde process reaction conditions to be 10%, hydrogen pressure 2 MPa, reaction temperature 373 K, and reaction time of 2 h. Under these conditions approximately 100% conversion of cinnamaldehyde toward hydrocinnamaldehyde was obtained for 10%Ni/SiO$_2$ prepared by a modified sol–gel method (i.e., very high activity and selectivity to the desired product only negligibly depended on the nickel precursor).

2.2.2 *Hydrogenation of Chloronitrobenzene*

Chemoselective hydrogenation of halogenated nitroaromatic compounds toward halogenated anilines, commonly known as a selective catalytic reduction process, is the most widely used method of synthesis of very important intermediates in production of analgesic and antipyretic drugs, dyes, pigments, and pesticides. And nickel is the most commonly used catalyst for this reaction.

Figure 2.4 shows different reaction pathways for *ortho*-chloronitrobenzene hydrogenation. The reaction has a rather complicated mechanism [19]. The main intermediate is *ortho*-chloronitrosobenzene, which disappeared when the conversion of *ortho*-chloronintrobenzene to aniline exceeded 99%,

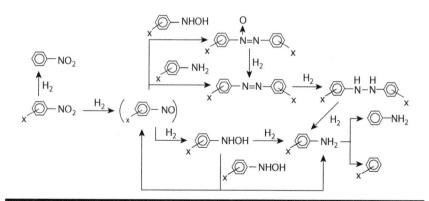

Figure 2.4   Reaction scheme of *ortho*-chloronitrobenzene hydrogenation. (Reproduced from J. Xiong, J. Chen, J. Zhang, *Catal. Commun.* 8 (2007) 345–350. With permission.)

became *ortho*-chloronitrosobenzene. The mechanism is even more complicated when starting with *para*-chloronitrobenzene (Figure 2.5) [20].

The reduction of *para*-chloronitrobenzene to *para*-chloroaniline involves the formation of nitroso- and hydroxylamine-intermediates (path I) that can participate in side reactions. The formation of dimeric azoxybenzene (path II) and azobenzene (path III) results from condensation–reduction involving *para*-chloroaniline, where conversion of the hydrazobenzene intermediate to *para*-chloroaniline (path IV) is favored in basic media [21]. Azoxybenzene/azobenzene formation, as highly toxic compounds, must be avoided in the clean production of *para*-chloroaniline. Disproportionation of the hydroxylamine intermediate to *p*-nitrosochlorobenzene and *para*-chloroaniline requires significant formation of *p*-(chorophenyl)-hydroxylamine (path V). Hydrogenolysis of the halo-substituent with subsequent formation of aniline (AN), benzene (BZ), and/or cyclohexane can occur either via the reduction of nitrobenzene (NB) (path VI) or as a result of the further conversion of *para*-chloroaniline (paths VII and VIII). Only the application of a chemoselective catalyst (Ni) in this process could radically limit the formation of undesired products and lead to formation of *p*-chloroaniline—the most desired product.

Catalytic properties of nickel catalysts in hydrogenation of chloronitrobenzenes are strongly related to the support used. The influence of the support on the activity and selectivity was investigated both in liquid and gas phases. Xiong et al. [19] demonstrated the effect of four supports (SiO_2, ZrO_2, TiO_2, and Al_2O_3) for nickel in hydrogenation of *ortho*-chloronitrobenzene in the liquid phase.

Table 2.1 contains catalytic data obtained for all of the catalysts. For Ni/γ-Al_2O_3, 100% of the selectivity to *ortho*-chloroaniline (desired product) was obtained; however, the catalytic activity of this material was very low. The best properties of the Ni/TiO_2 catalyst (99.9% conversion of *o*-chloronitrobenzene and 99.5% selectivity of *o*-chloroaniline at 1.5 MPa and 363 K) were

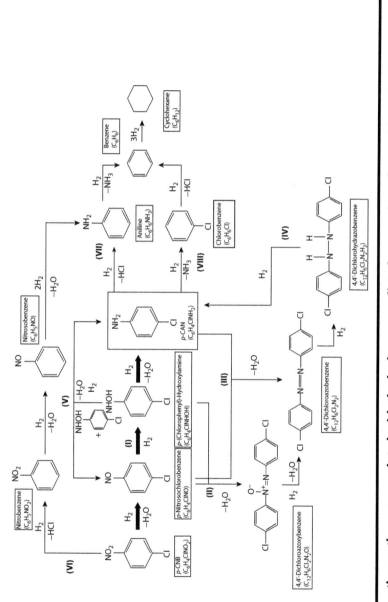

Figure 2.5 **Reaction pathways associated with the hydrogen mediated conversion of *para*-chloronitrobenzene (*p*-CNB). The targeted route (I) to *para*-chloroaniline (*p*-CAN) is represented by bold arrows. (Reproduced from F. Cárdenas-Lizana, S. Gómez-Quero, M. A. Keane, *Appl. Catal. A* 334 (2008) 199–206. With permission.)**

Table 2.1 Catalytic Performance of Nickel-Based Catalysts with Different Supports for the Hydrogenation of *Ortho*-Chloronitrobenzene

Catalyst	Reaction Time (min)	Turnover Frequency (s^{-1})	Ortho-Chloronitrobenzene Conversion (%)	Ortho-Chloroaniline Selectivity (%)
Ni/γ-Al$_2$O$_3$	30		0.6	100
	120		1.2	100
Ni/SiO$_2$	30	0.2	19.3	93.7
	660		60.8	96.4
Ni/ZrO$_2$	30	1.4	26.4	89.1
	240		99.6	98.1
Ni/TiO$_2$	30	34.9	99.1	99.2
	60		99.9	99.6

Source: Reproduced from J. Xiong, J. Chen, J. Zhang, *Catal. Commun.* 8 (2007) 345–350. With permission.

perhaps attributed to the strong polarization of the N=O band induced by oxygen vacancies of TiO_x, which was produced by a high-temperature reduction. The N=O band polarized was attacked easily by hydrogen dissociatively adsorbed on the nickel particles. The overall activity of all nickel catalysts changed in order: $Ni/TiO_2 > Ni/ZrO_2 > Ni/SiO_2 > Ni/Al_2O_3$.

Very different activity of nickel catalysts has been determined in the hydrogenation of *para*-chloronitrobenzene. Authors showed the highest activity with 100% of selectivity to *para*-chloroaniline for alumina-supported nickel [20], and the activity sequence could be present as $Ni/Al_2O_3 > Ni/SiO_2$ > Ni/activated carbon > Ni/graphite. Therefore, it is clearly evident that catalytic behavior of the nickel/support system also depends on the other factors, such as nickel particle size distribution, kind of metal precursor used, or the method of catalyst preparation and reaction conditions. Figure 2.6 shows the representative TEM results obtained for Ni/Al_2O_3 (a), Ni/ activated carbon (b), and Ni/graphite (c). The comparative study between these catalysts showed the smallest nickel particle size in the case of Ni/Al_2O_3, which supposedly leads to stronger Ni/support interactions. On the other hand, hydrogenation of *para*-chloronitrobenzene at 393 K overall supported Ni-generated *para*-chloroaniline as the only product with no evidence of dechlorination or ring reduction. This means that nickel particle size did not affect the selectivity toward the desired product *para*-chloroaniline.

Generally, carbon materials (in the form of powder, nanofibers [22], nanowires [23], expanded graphite [24], or as filaments [25]) play the crucial role as the support for highly active and stable nickel in hydrogenation of chloronitrobenzene with >99% selectivity to chloroaniline. Recently, very interesting results were obtained for hydrogenation of *ortho*-chloronitrobenzene by nickel carbide-promoted nickel/carbon nanofiber nanocomposite catalysts (NiC-Ni/CNFs). The uniqueness of this catalyst consists in its generation by an in situ facile catalytic chemical vapor deposition methodology (CCVD) [26].

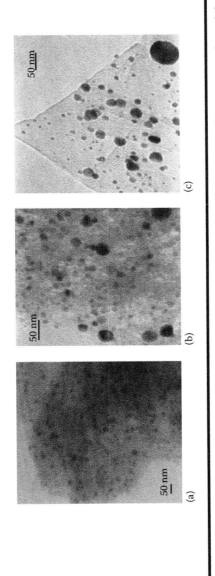

Figure 2.6 Representative TEM images of (a) Ni/Al$_2$O$_3$, (b) Ni/AC, and (c) Ni/graphite. (Reproduced from F. Cárdenas-Lizana, S. Gómez-Quero, M. A. Keane, *Appl. Catal. A* 334 (2008) 199–206. With permission.)

In the NiC-Ni/CNFs system, surprisingly, physically isolated metallic Ni NPs (nanoparticles) coated by graphitic layers are catalytically active for hydrogenation, even though H₂ and *ortho*-chloronitrobenzene probably are not in direct contact with active sites. It is found that as-formed NiC-Ni/CNFs catalysts present very high selectivity to *ortho*-chloroaniline (>98.0%). Therefore, Ni NPs probably interact with graphitic carbon shells and affect the properties of the outer walls. As a result, the outstanding activity of NiC-Ni/CNFs catalysts should be related to the interactions between metallic Ni nanoparticles formed and graphitic carbon shells, which change their local work function. In a way, as for NiC-Ni/CNFs catalysts, the high surface area of CNFs in situ that is formed may facilitate the adsorption of hydrogen molecules on the surface of CNFs in the course of hydrogenation, and thus H₂ adjacent to active sites is easy to be dissociated. The schematic form demonstrated the mechanism of the reaction (Figure 2.7). Authors have shown that the coexistence of pure nickel particles and NiC$_x$ forms led to very active, stable, and selective formation to chloroaniline catalysts.

As was shown before, nickel particles are usually supported on 1D and 2D materials, such as different forms of carbons

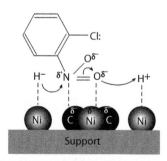

Figure 2.7 Schematic representation of hydrogenation mechanism of *o*-CNB over the NiC-Ni/CNFs catalyst. (Reproduced from J. Kang, R. Yan, J. Wang, L. Yang, G. Fan, F. Li, *Chem. Eng. J.* 275 (2015) 36–44. With permission.)

or oxides. An attractive alternative for these kinds of materials could be montmorillonite clay [27]. Figure 2.8 shows the TEM images of a montmorillonite-supported nickel catalyst. Application of environmentally friendly, cheap, easily available, and robust montmorillonite as the support for approximately 5 nm nickel nanoparticles (Figure 2.8) led to formation of an

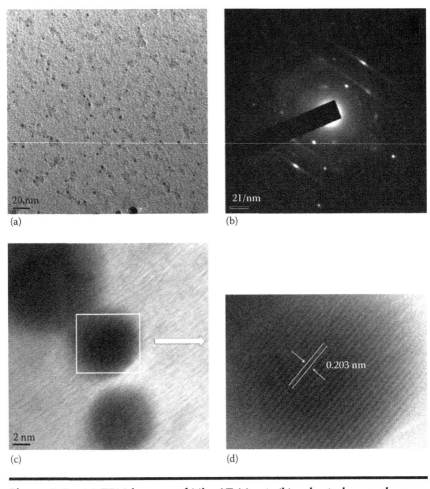

Figure 2.8 **(a) TEM images of Nio-AT-Mont; (b) selected area electron diffraction (SEAD) pattern of Nio-AT-Mont; (c) HR-TEM images of Nio-AT-Mont; and (d) enlarged image of HR-TEM with fringe spacing. (Reproduced from D. Dutta, D. K. Dutta, *Appl. Catal. A* 487 (2014) 158–164. With permission.)**

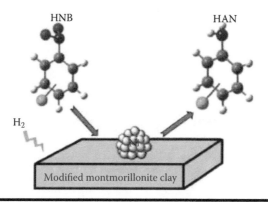

Figure 2.9 **Schematic representation of the selective hydrogenation of halonitrobenzene (HNB) to corresponding haloaniline (HAN) over Ni supported on montmorillonite clay. (Reproduced from D. Dutta, D. K. Dutta, *Appl. Catal. A* 487 (2014) 158–164. With permission.)**

efficient and selective heterogeneous catalyst for hydrogenation of halonitrobenzene (HNB) to corresponding haloaniline (HAN) with conversion 78%–100% and selectivity 96%–99.4% (Figure 2.9).

Furthermore, this metal nanocatalyst could be recycled and reused several times without significant loss of catalytic activities. Moreover, TEM investigations of recovered catalysts after two or three runs (Figure 2.10) did not show any spectacular sintering of the metallic phase.

2.2.3 *Hydrogenation of Benzene to Cyclohexane*

The hydrogenation of aromatics is of major importance in the chemical industry because of the stringent environmental regulations governing their concentration in diesel fuels [28]. In the petrochemical industry or in polymer production (the first step of polyamide synthesis), cyclohexane is mainly produced from the hydrogenation of benzene, according to the reaction

$$C_6H_6 + 3H_2 \leftrightarrow C_6H_{12}$$

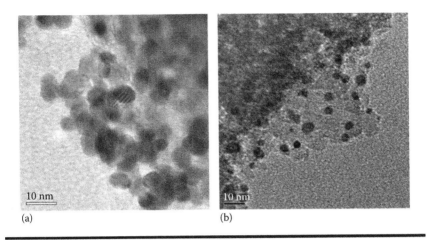

Figure 2.10 **HR-TEM image of recovered catalyst after (a) second run and (b) third run (Ni⁰ nanoparticles supported on AT-Mont). (Reproduced from D. Dutta, D. K. Dutta, *Appl. Catal. A* 487 (2014) 158–164. With permission.)**

Cyclohexane can also be recovered directly from natural gasoline and petroleum naphtha [29]. However, the dehydrogenation of cyclohexane and its derivatives (present in naphthas) to aromatic hydrocarbons is an important reaction in the catalytic reforming process used in refineries to produce high-octane motor fuels [30]. From petroleum fractions, it is difficult to obtain the very high quality cyclohexane product achieved by benzene hydrogenation [31]. Additionally, benzene hydrogenation has been chosen as a model aromatic substance [32]. This reaction has also been used as model reaction in heterogeneous catalysis by metals where metal–support interactions are involved [33–39], and the desired product of benzene hydrogenation—cyclohexane—is an important chemical intermediate for the synthesis of ε-caprolactone, caprolactam, nylon-66, and nylon 6. Among the plethora of catalysts shown to be active in hydrogenation of benzene toward cyclohexane, nickel plays a major role. The choice of nickel is mainly due to its lower cost compared to noble metals. Nickel supported on various oxide catalysts is currently used in industry [40].

Benzene hydrogenation can be considered as a structure-sensitive reaction and the rate of this reaction can be related to the unit area of metal. For poorly dispersed nickel/silica samples obtained by reduction at high temperature, the decrease in the intrinsic activity has been attributed to a strong metal–support interaction (SMSI) [41]. In the case of highly dispersed materials, the changes in the catalytic activity were related to the combined effects of particle size, surface coverage with adsorbed species, and active site dimension.

It was concluded that the specific activity of reduced nickel for benzene hydrogenation decreases with the increasing acid activity of the support in the order $SiO_2 > Al_2O_3 >$ silica–alumina > Y zeolites, but these activities remain constant in the faujasite acidity range from NaY to MgY zeolites [42]. The negative effect of acidity was explained by the formation of carbonaceous compounds in the cavities, at channel intersections, or on outer surfaces of catalysts containing acidic sites. The constant activity of nickel-loaded Y zeolites could be the effect of the passivation of strong acid sites accomplished by a treatment with alkaline solution.

On the other hand, the application of "inert" carbon nanotubes (CNTs) showed the importance of the location of nickel active sites in hydrogenation of benzene [43]. Authors showed that depending on the preparation method, nickel could be located inside nanotubes (Ni-filled CNTs) or deposited on CNTs, as presented in Figure 2.11.

The catalytic activity of Ni filled inside CNTs was 4.6 times higher than that of Ni deposited outside CNTs in hydrogenation of benzene (Figure 2.12). Authors explained that the enhanced catalytic activity can be attributed to the confinement of CNTs with more defects, which provides facile reduction, reinforced reactivity, and increased reactant concentrations due to a larger charge transfer and deficient electrons in the tubular microreactor. And the gaps formed on the sidewall of CNTs during the treatment process also played an important role for decreasing the diffuse resistance kinetically [43].

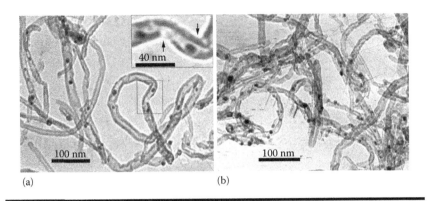

(a)   (b)

Figure 2.11   TEM micrograph of the catalysts. (a) Ni-filled CNTs (inset shows the amplification of the square part) and (b) Ni-deposited CNTs. (Reproduced from H. Yang, S. Song, R. Rao, X. Wang, Q. Yu, A. Zhang, *J. Mol. Catal.* 323 (2010) 33–39. With permission.)

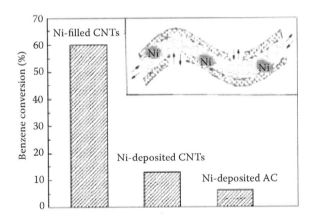

Figure 2.12   Benzene conversion on Ni-CNTs. (Reproduced from H. Yang, S. Song, R. Rao, X. Wang, Q. Yu, A. Zhang, *J. Mol. Catal.* 323 (2010) 33–39. With permission.)

Boudjahem, Bouderbala, and Bettahar [44] also showed the influence of preparation method on catalytic behavior of Ni/ SiO_2 catalysts. Both nickel catalysts were prepared by impregnation of silica by nickel acetate and then separated on the two portions. One of the portions was treated with 24%–26% aqueous hydrazine, and the second one was calcined in air. Although both materials were reduced in the same way before kinetic run, they showed different catalytic behavior in hydrogenation of benzene to cyclohexane. Figure 2.13 shows the results obtained for Ni/SiO_2 prepared by conventional (marks as AC1) and unconventional—with hydrazine (marks as Ni)—methods.

Nickel catalysts prepared by a nonclassical method showed much better catalytic activity than Ni/SiO_2 prepared by the classical method. This phenomenon was explained by higher nickel dispersion obtained for the catalyst prepared using the hydrazine.

A highly thermal, stable, metal–organic framework (MOF) containing nickel, Ni/MIL-120, exhibited excellent catalytic activities in comparison with a classic Ni/Al_2O_3 catalyst in benzene conversion toward cyclohexane. The authors explained the very high activity of these catalysts as the effect of weaker interaction between the Ni species and the MOF support than in the case of typical Ni/support systems [45].

Wojcieszak et al. [46] showed that the intensity of hydrogen spillover strongly decreased in the presence of Al in nickel

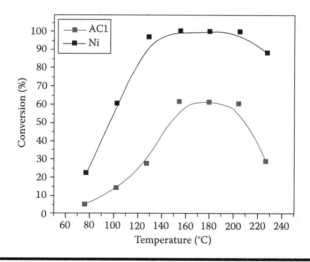

Figure 2.13 Activity in benzene hydrogenation of the Ni nonclassical and classical AC1 catalysts. (Reproduced from A.-G. Boudjahem, W. Bouderbala, M. Bettahar, *Fuel Process. Technol.* 92 (2011) 500–506. With permission.)

containing mesoporous molecular sieves MCM-41. The kinetic study with the mechanical mixtures suggested changes in the surface reactions due to the contribution of the support to the chemical processes and the fact that the migration of hydrogen atoms from the metal phase onto the support oxide was assisted by the aromatic hydrocarbon.

Recently, special attention has been paid to bimetallic catalysts in hydrogenation of benzene to cyclohexane [47,48]. Among the different dopants, the most attention was paid to Ni modified by Pd, Ru, Pt, and Cu. While the addition of Pd, Pt, and Ru [49] to nickel beneficially affected the activity of nickel catalysts, the addition of copper decreased the nickel phase dispersion as well as the conversion, whereas it increased the formation of carbon deposits during benzene hydrogenation. Very interesting results for hydrogenation of benzene were obtained for a nontypical Ni/Mo/N catalyst with a structure presented in Figures 2.14 and 2.15 [50]. A Ni_2Mo_3N catalyst, prepared by the decomposition of the HMT complex, gave 100% cyclohexane in benzene hydrogenation with activity higher than a typical Ni_3C system [50]. On the other hand, Ru/Al_2O_3-ZrO_2-NiO/cordierite also showed high and stable conversion in the benzene hydrogenation process [51].

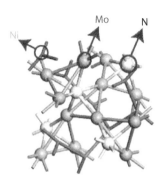

Figure 2.14 **Structure of the Ni_2Mo_3N catalyst. (Reproduced from C. Qi, F. Jie, L. Wenying, X. Kechang, *Chin. J. Catal.* 34 (2013) 159–166. With permission.)**

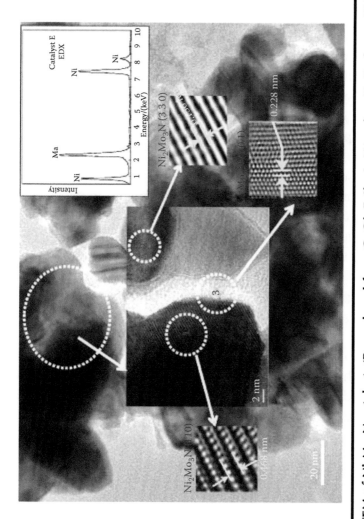

Figure 2.15 TEM of Ni₂Mo₃N catalyst. (Reproduced from C. Qi, F. Jie, L. Wenying, X. Kechang, *Chin. J. Catal.* **34** (2013) 159–166. With permission.)

2.3 Hydrotreatment by Nickel

2.3.1 Hydrodenitrogenation and Hydrodesulfurization

Hydrodenitrogenation (HDN) and hydrodesulfurization (HDS) are the processes important in the technological point of view. Hydrodenitrogenation reduces the nitrogen content of a petroleum stream, and HDS is a catalytic process widely used to remove sulfur (S) from natural gas and from refined petroleum products such as gasoline or petrol. The general idea of HDS and HDN is presented in Figure 2.16.

These processes are significant because

a. The sulfur- and nitrogen-containing impurities in hydrocarbon fuels have a severe environmental impact, resulting from the contribution of sulfur and nitrogen oxides (produced during combustion) to acid rain.
b. The sulfur- and nitrogen-containing impurities are effective catalyst poisons that prevent crude feedstock from being used for subsequent chemical transformations.

Due to its properties and low cost, nickel is the attractive alternative for noble metals in hydrotreatment reactions.

From the eighties of the twentieth century, nickel has been considered as metal with the potential to be an efficient catalyst for the HDS process [52]. It should be mentioned here

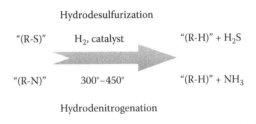

Figure 2.16 Concept of hydrodenitrogenation and hydrodesulfurization processes.

that HDN and HDS are usually run together. Although few works concern monometallic nickel in these processes, interest in bimetallic systems with nickel as dopant is systematically growing. Very interesting results for nickel modified by boron in HDN/HDS reactions were obtained by Lewandowski [53], who showed that the HDN reaction took place mainly on the metallic phase (Ni^0). During the HDN/HDS reaction the activity of the catalyst was attributed mainly by the presence of Ni^0 and/or Ni$_3$B phases. HDS catalysts are traditionally composed of Mo or W, but they are efficient only after modification by Ni or Co. The supported (Ni/Co) Mo catalysts are still predominantly used in most refineries [54,55].

On the other hand, Trejo et al. [56] showed that, independently of used support, nickel-containing catalysts exhibited the highest activity in HDS of benzothiophene. The activity of investigated catalysts decreased in the order NiMo > NiW > CoMo over alumina–zeolite, while for the support effect the order was Al$_2$O$_3$–zeolite > Al$_2$O$_3$–TiO$_2$ > Al$_2$O$_3$–SiO$_2$. The key role of nickel has also been demonstrated in HDS on a garnierite–nickel ore mineral [57]. Leaching of nickel from molybdenum containing garnierite led to decreasing of catalyst activity. The beneficial role of nickel has been demonstrated for the HDN process also [58]. The enhancement in the HDN and HDS activity with nickel addition has been attributed to the improvement in the reducibility of molybdenum and the formation of the Ni-Mo-O phase [58]. Other authors [59] have explained its role in bimetallic systems due to the enhanced hydrogenation ability of nickel.

2.3.2 Hydrodechlorination

In the earlier parts of this chapter the role of nickel in the reactions important in the technological point of view has been indicated. As one of the most universal resource-saving methods for reprocessing chloro-organic compounds, hydrodechlorination (HDC) is important, especially from an ecological point of view. This reaction could be effectively catalyzed by palladium and platinum [60] or noble metals doped by transition metals [61–63], but the very high cost of noble metals significantly reduces their practical application. An attractive alternative for Pd and Pt catalysts is less expensive nickel [64].

The advantage of nickel-based catalysts is a very high selectivity toward unsaturated products such as ethylene in 1,2-dichloroethane hydrodechlorination [65,66] or benzene in HDC of chlorobenzene [67]. Catalytic behavior of nickel catalysts in the HDC process strongly depends on the used support, nickel dispersion, and reaction conditions [64–67].

Active carbons, due to their high adsorption capacity for organic compounds and high stability under the aggressive conditions of the reaction, play the most important role as the supports for nickel catalysts [68]. Hydrodechlorination of chloro-organic compounds on Ni/active carbon materials gives fully dechlorinated, but usually unsaturated, hydrocarbons. Recently, there has been growing interest in nickel containing zeolites, too [66]. Depending on the used zeolites, different products were observed—for example, application of the support with a predominance of Brønsted acidic sites led to the formation of not fully dechlorinated products, like vinyl chloride in 1,2-dichloroethane hydrotreatment [66].

2.4 Preparation of Nickel Nanoparticles

Independently of the kind of catalytic reaction and its conditions, homogeneity of the catalyst plays the crucial role in the heterogeneous catalysis, especially in the case of structural sensitive processes. Application of different methods of synthesis of metal(s) nanoparticles could affect the catalytic behavior of these materials. Therefore, it seems important to present some of the efficient methods of nickel nanoparticle preparation leading to obtaining monodisperse materials in relation to their catalytic activity.

2.4.1 Synthesis of Unsupported Ni Nanoparticles

Many studies suggest the crucial role of nickel nanoparticles' size in heterogeneous catalysis. Therefore, particular attention will be given to the preparation methods that lead to formation of well-defined nickel nanoparticles. However, finding the procedure, which prohibits the agglomeration and guarantees the formation of nanoparticles with uniform size and morphology, is rather problematic.

In recent years different preparation methods have been reported, such as sol–gel [69], chemical reduction [70,71], microemulsion [72], thermal decomposition of organometallic precursor [73,74], electrodeposition [75], the polyol method [76], hydrothermal or solvothermal methods [77], sonochemical reduction [78], two-step postsynthesis or microwave synthesis [79]. Among all of these methods, the chemical reduction is considered to be the most convenient one.

Wang et al. [80] synthesized phase-pure nickel nanoparticles via a simple chemical reduction route in anhydrous ethanol using citric acid, cetyltrimethylammonium bromide, D-sorbitol, and Tween 40 as organic modifiers, which they found active in the hydrogenation of *p*-nitrophenol to *p*-aminophenol. The average particle size depends on the organic modifier used—that is, citric acid (9 nm) > CTAB (64 nm) > Tween 40 (297 nm) > D-sorbitol (339 nm) (Figure 2.17). Additionally, the modifiers also yielded different morphologies. It was shown that citric acid favors the formation of small nickel nanoparticles with point defects. On the other hand, the application of CTAB as an organic modifier led to formation of nickel nanoparticles with superimposed periodic structure. The presence of Tween 40 has given nickel nanoparticles with a relatively perfect internal structure. The presence of D-sorbitol as an organic modifier led to the formation of nickel nanoparticles with twinned and superimposed periodic structures.

According to Wang et al. [80], the catalytic activity of nickel nanoparticles depends on the nickel particles' size (increased

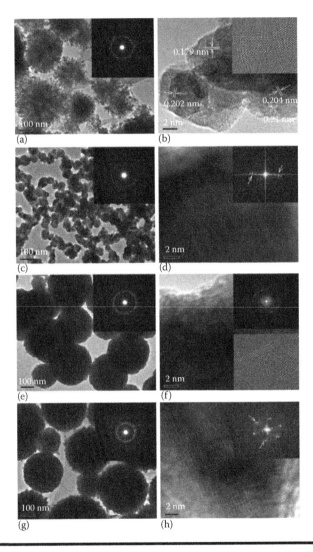

Figure 2.17 TEM and HRTEM images of the ethanol-washed nickel nanoparticles prepared by reducing nickel oxalate with hydrazine hydrate and using (a, b) citric acid, (c, d) CTAB, (e, f) Tween 40, and (g, h) ᴅ-sorbitol as organic modifiers, respectively. The insets in the TEM images show the corresponding SAED patterns, and the diffraction rings are assigned to {111}, {200}, {220}, and {311} reflections of fcc nickel metal. The insets in the HRTEM images show the corresponding FFT and inverse FFT images. (Reproduced from A. Wang, H. Yin, H. Lu, J. Xue, M. Ren, T. Jiang, *Langmuir* 25 (2009) 12736–12741. With permission.)

with decreasing of their size) and the crystal structure (the nickel nanoparticles with internal planar defects had higher catalytic activity than those with perfect internal structure). But, generally, independently of the kind of modifier, all of the investigated materials showed higher catalytic activity and selectivity than the Raney Ni commercial catalysts in the hydrogenation of *p*-nitrophenol to *p*-aminophenol.

Pan et al. [81] prepared monodisperse nickel nanoparticles with different sizes via the thermal decomposition method, using nickel acetylacetonate as precursor and trioctylphosphine as surfactant in oleylamine. By changing the process conditions during organometallic precursor thermal decomposition, the size and morphology of nickel nanoparticles could be well controlled. The presence of the trioctylphosphine as surfactant was crucial to prevent particle agglomeration, as well as for controlling the size and morphology of nanoparticles. The as-synthesized nickel nanoparticles exhibited monodisperse sphere-like morphology and had different size distribution with the changing of the P:Ni precursor ratio. The typically spherical shape indicated that the growth of nickel nanoparticles was isotropic. Decreasing the molar ratio of the P:Ni precursor resulted in an increase in the size of nanoparticles. These results indicated that trioctylphosphine can stabilize the particles by absorbing the surface of nickel nanoparticles and limiting the aggregation. However, increasing the molar ratio of P:Ni precursor from 2:1 to 4:1, the size of as-synthesized nickel NPs was not changed, which indicated that the adsorption of trioctylphosphine reached the saturation state.

2.4.2 Synthesis of Supported Ni Nanoparticles

The most popular and the easiest method to prepare Ni/support systems is the incipient wetness impregnation [82]. However, its main disadvantage is the formation of non-homogenous dispersed nickel species with different nickel particles size. The two-step postsynthesis method is a very promising method leading to formation of very small and stable nickel nanoparticles. However, it is especially dedicated for zeolites like BEA. Figure 2.18 shows the probable pathway of formation of pseudotetrahedral Ni^{2+} in nickel containing beta zeolite [83].

According to this method tetraethylammonium BEA (TEABEA) (Si/Al = 17) zeolite, after calcination to remove the template, was treated in concentrated nitric acid to obtain a dealuminated form of zeolite with Si:Al ratio > 1300. Afterward, the zeolites with free T-atom vacant sites were impregnated by nickel solution to obtained pseudotetrahedral nickel species. Then, nickel-containing zeolite was calcinated and reduced to 873 K to obtain excellent dispersal and uniquely small Ni^0 nanoparticles. Authors demonstrated that nickel catalysts prepared this way are active and stable in hydrogenation of chloro-organic compounds to hydrocarbons, with negligible effect of sintering during catalytic reaction. Figure 2.19 contains STEM and HRTEM images for reduced and spent catalysts. The catalysts were successfully tested in the hydrodechlorination of 1,2-dichloroethane (Figure 2.20) [84].

2.5 Conclusions and Prospects for Application of Nickel Catalysts

Nickel as a heterogeneous catalyst exhibits excellent properties in the processes important from both technological and ecological points of view. Ni/support systems find applications in hydrogenation of C=C double bonds, chemoselective hydrogenation of α,β-unsaturated aldehydes to unsaturated alcohols, and hydrogenation of nitro groups in the presence of other reactive bonds. The advantage of nickel catalysts is their low cost and a very high selectivity toward desired products. Catalytic behavior of Ni-containing materials depends on the

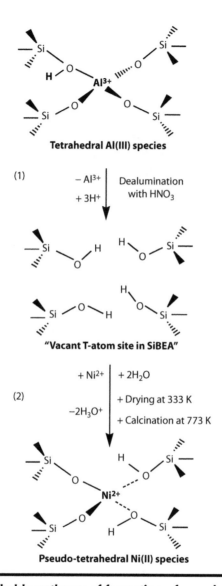

Figure 2.18 Probable pathway of formation of pseudo-tetrahedral Ni^{2+} in C-Ni$_x$SiBEA zeolites. (Reproduced from R. Baran, I. I. Kamińska, A. Śrębowata, S. Dzwigaj, *Micro. Meso. Mater.* 169 (2013) 120–127. With permission.)

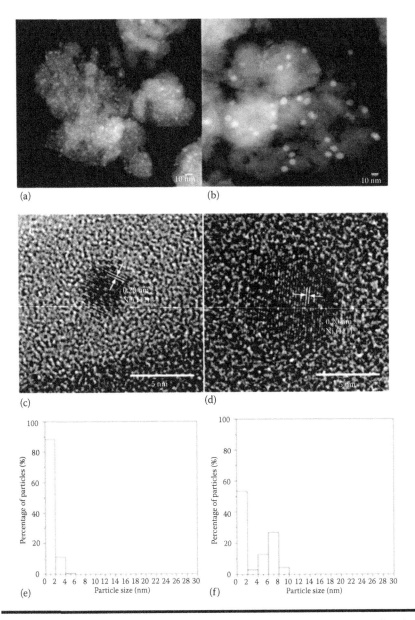

Figure 2.19 STEM and HRTEM images and nickel particle size distribution of red-C-Ni2.0SiBEA (a, c, and e) and spent red-C-Ni2.0SiBEA (b, d, and f) effect of sintering. (Reproduced from A. Śrębowata, R. Baran, D. Lomot, D. Lisovytskiy, T. Onfroy, S. Dzwigaj, *Appl. Catal. B* **147** (2014) 208–220. With permission.)

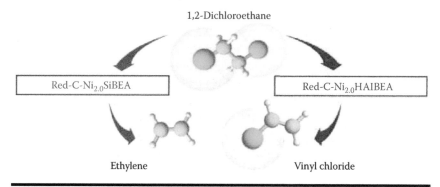

1,2-Dichloroethane

Red-C-Ni$_{2.0}$SiBEA

Red-C-Ni$_{2.0}$HAlBEA

Ethylene

Vinyl chloride

Figure 2.20   Schematic representation of hydrodechlorination process of 1,2-dichloroethane on Ni supported on zeolites. (Reproduced from A. Śrębowata, R. Baran, D. Lomot, D. Lisovytskiy, T. Onfroy, S. Dzwigaj, *Appl. Catal. B* 147 (2014) 208–220. With permission.)

metal dispersion, kind of support, and preparation method. Hydrogenation catalyzed by nickel provides a basis for sustainable, innovative, and efficient industrial technology.

References

1. S. Z. Tasker, E. A. Standley, T. F. Jamison, Recent advances in homogeneous nickel catalysis. *Nature* 509 (2014) 299–309.
2. R. L. Augustine, *Heterogeneous catalysts in organic synthesis*, Marcel Dekker, New York (1995).
3. M. S. Ide, B. Hao, M. Neurock, R. J. Davis, Mechanistic insights on the hydrogenation of α,β-unsaturated ketones and aldehydes to unsaturated alcohols over metal catalysts. *ACS Catal.* 2 (2012) 671–683.
4. G. C. Torres, S. D. Ledesma, E. L. Jablonski, S. R. De Miguel, O. A. Scelza, Hydrogenation of carvone on Pt–Sn/Al$_2$O$_3$ catalysts. *Catal. Today* 48 (1999) 65–72.
5. Q. Luo, T. Wang, M. Beller, H. Jiao, Acrolein Hydrogenation on Ni(111). *J. Phys. Chem. C* 117 (2013) 12715–12724.
6. B. Coq, F. Figueras, P. Geneste, C. Moreau, P. Moreau, M. Warawdekar, Hydrogenation of α,β-unsaturated carbonyls: Acrolein hydrogenation on Group VIII metal catalysts. *J. Mol. Catal.* 78 (1993) 211–226.

7. M. L. Toebes, Y. H. Zhang, J. Hájek, T. A. Nijhuis, J. H. Bitter, A. J. van Dillen, D. Y. Murzin, D. C. Koningsberger, K. P. de Jong, Support effects in the hydrogenation of cinnamaldehyde over carbon nanofiber-supported platinum catalysts: Characterization and catalysis. *J. Catal.* 226 (2004) 215.

8. C. Milone, C. Crisafulli, R. Ingoglia, L. Schipilliti, S. Galvagno, A comparative study on the selective hydrogenation of α,β unsaturated aldehyde and ketone to unsaturated alcohols on Au supported catalysts. *Catal. Today* 122 (2007) 341.

9. M. Lashdaf, M. Tiitta, T. Venalainen, H. Osterholm, A. O. I. Krause, Ruthenium on Beta Zeolite in Cinnamaldehyde Hydrogenation. *Catal. Lett.* 94 (2004) 7.

10. J. P. Tessonnier, L. Pesant, G. Ehret, M.J Ledoux, C. P.-Huu, Pd nanoparticles introduced inside multi-walled carbon nanotubes for selective hydrogenation of cinnamaldehyde into hydrocin-namaldehyde. *Appl. Catal. A* 288 (2005) 203.

11. B. Viswanathan, K. R. Krishnamurthy, R. Mahalakshmi, M. G. Prakash, Selective hydrogenation of cinnamaldehyde on nickel nano particles supported on titania—Role of catalyst preparation methods. *Catal. Sci. Technol.* 5 (2015) 3313–3321.

12. C. Rudolf, B. Dragoi, A. Ungureanu, A. Chirieac, S. Royer, A. Nastro, E. Dumitriu, NiAl and CoAl materials derived from takovite-like LDHs and related structures as efficient chemoselec-tive hydrogenation catalysts. *Catal. Sci. Technol.* 4 (2014) 179–189.

13. K.-Y. Jao, K.-W. Liu, Y.-H. Yang, A.-N. Ko, Vapour Phase Hydrogenation of Cinnamaldehyde over Nano Ni and Ni/SiMCM-41 Catalysts. *J. Chin. Chem. Soc.* 56 (2009) 885–890.

14. S.-Y. Chin, F.-J. Lin, A.-N. Ko, Vapour phase hydrogenation of cinnamaldehyde over Ni/γ-Al$_2$O$_3$ catalysts: Interesting reaction network. *Catal. Lett.* 132 (2009) 389–394.

15. S. Gryglewicz, A. Śliwak, J. Ćwikła, G. Gryglewicz, Performance of carbon nanofiber and activated carbon supported nickel catalysts for liquid-phase hydrogenation of cinnamaldehyde into hydrocinnamaldehyde. *Catal. Lett.* 144 (2014) 62–69.

16. L. J. Malobela, J. Heveling, W. G. Augustyn, L. M. Cele, Nickel–Cobalt on carbonaceous supports for the selective catalytic hydrogenation of cinnamaldehyde. *Ind. Eng. Chem. Res.* 53 (2014) 13910–13919.

17. S. Y. Chin, L. Fang-Jen, A-N. Ko, Vapour phase hydrogenation of cinnamaldehyde over Ni/γ-Al$_2$O$_3$ catalysts: Interesting reaction network. *Catal. Let.* 132 (2009) 389–394.

18. Q. Fan, Y. Liu, Y. Zheng, W. Yan, Preparation of Ni/SiO$_2$ catalyst in ionic liquids for hydrogenation. *Front. Chem. Eng. in China* 2 (2008) 63–68.
19. J. Xiong, J. Chen, J. Zhang, Liquid-phase hydrogenation of *o*-chloronitrobenzene over supported nickel catalysts. *Catal. Commun.* 8 (2007) 345–350.
20. F. Cárdenas-Lizana, S. Gómez-Quero, M. A. Keane, Clean production of chloroanilines by selective gas phase hydrogenation over supported Ni catalysts. *Appl. Catal. A* 334 (2008) 199–206.
21. I. A. Ilchenko, A. V. Bulatov, I. E. Uflyand, V. N. Sheinker, Hydrogenation of chloronitrobenzenes on heterogenized Pd(II) chelates. *Kinet. Catal.* 32 (1991) 691.
22. C. Wang, J. Qiu, C. Liang, L. Xing, X. Yang, Carbon nanofiber supported Ni catalysts for the hydrogenation of chloronitrobenzenes. *Catal. Commun.* 9 (2008) 1749–1753.
23. H. Li, H. Lin, S. Xie, W. Dai, M. Qiao, Y. Lu, H. Li, Ordered mesoporous Ni nanowires with enhanced hydrogenation activity prepared by electroless plating on functionalized SBA-15. *Chem. Mater.* 20 (2008) 3936–3943.
24. W. Li, C. Han, W. Liu, M. Zhang, K. Tao, Expanded graphite applied in the catalytic process as a catalyst support. *Catal. Today* 125 (2007) 278–281.
25. N. Mahata, A. F. Cunha, J. J. M. Órfão, J. L. Figueiredo, Hydrogenation of chloronitrobenzenes over filamentous carbon stabilized nickel nanoparticles. *Catal. Commun.* 10 (2009) 1203–1206.
26. J. Kang. R. Yan, J. Wang, L. Yang, G. Fan, F. Li, In situ synthesis of nickel carbide-promoted nickel/carbon nanofibers nanocomposite catalysts for catalytic applications. *Chem. Eng. J.* 275 (2015) 36–44.
27. D. Dutta, D. K. Dutta, Selective and efficient hydrogenation of halonitrobenzene catalyzed by clay supported Ni^0-nanoparticles. *Appl. Catal. A* 487 (2014) 158–164.
28. R. Wojcieszak, S. Monteverdi, M. Mercy, I. Nowak, M. Ziolek, M. M. Bettahar, Nickel containing MCM-41 and AlMCM-41 mesoporous molecular sieves: Characteristics and activity in the hydrogenation of benzene. *Appl. Catal. A* 268 (2004) 241–253.
29. S. Matar, L. F. Hatch, *Chemistry of petrochemical processes,* 2nd ed., Butterworth-Heinemann, Boston (2001) pp. 281–283.
30. N. Dachos, A. Kelly, D. Felch, E. Reis. In *Handbook of petroleum refining processes, part 4,* 2nd ed., R. A. Meyers (ed.), McGraw–Hill (1997) pp. 4.1–4.26.

31. J. Carrero-Mantilla, M. Llano-Restrepo, Vapor-phase chemical equilibrium for the hydrogenation of benzene to cyclohexane from reaction-ensemble molecular simulation. *Fluid Phase Equilibria* 219 (2004) 181–193.

32. A. Boudjahem, S. Monteverdi, M. Mercy, G. Ghanbaja, M. M. Bettahar, Nickel nanoparticles supported on silica of low surface area. Hydrogen chemisorption and TPD and catalytic properties. *Catal. Lett.* 84 (2002) 115–122.

33. R. Molina, G. Poncelet, Hydrogenation of benzene over alumina-supported nickel catalysts prepared from Ni(II) acetylacetonate. *J. Catal.* 199 (2001) 162.

34. B. Coughlan, M. A. Keane, The hydrogenation of benzene over nickelsupported Y zeolites. Part 1. A kinetic approach. *Zeolites* 11 (1991) 12.

35. K. J. Yoon, M. A. Vannice, Benzene hydrogenation over iron: II. Reaction model over unsupported and supported catalysts. *J. Catal.* 82 (1983) 457.

36. M. A. Keane, The Hydrogenation ofo-,m-, andp-Xylene over Ni/SiO₂. *J. Catal.* 166 (1997) 347.

37. A. Boudjahem, S. Monteverdi, M. Mercy, G. Ghanbaja, M. M. Bettahar, Nickel nanoparticles supported on silica of low surface area. Hydrogen chemisorption and TPD and catalytic properties. *Catal. Lett.* 84 (2002) 115–122.

38. D. Franquin, S. Monteverdi, S. Molina, M. M. Bettahar, Y. Fort, Colloidal nanometric particles of nickel deposited on γ-alumina: Characteristics and catalytic properties. *J. Mater. Sci.* 34 (1999) 4481.

39. S. Lefondeur, S. Monteverdi, S. Molina, M. Bettahar, Y. Fort, Nickel nanoparticles inserted in tBuONa matrix deposited on alumina Part II Thermal treatment and nickel content effects on their stabilty and catalytitic activity. *J. Mater. Sci.* 36 (2001) 2633.

40. P. G. Savva, K. Goundani, J. Vakros, K. Bourikas, Ch. Fountzoula, D. Vattis, A. Lycourghiotis, Ch. Kordulis, Benzene hydrogenation over Ni/Al₂O₃ catalysts prepared by conventional and sol–gel techniques. *Appl. Catal. B* 79 (2008) 199–207.

41. G. A. Martin, J. A. Dalmon, Benzene hydrogenation over nickel catalysts at low and high temperatures: Structure-sensitivity and copper alloying effects. *J. Catal.* 75 (1982) 233.

42. A. Lewandowska, S. Monteverdi, M. Bettahar, M. Ziolek, MCM-41 mesoporous molecular sieves supported nickel—physico-chemical properties and catalytic activity in hydrogenation of benzene. *J. Mol. Catal.* A 188 (2002) 85–95.

43. H. Yang, S. Song, R. Rao, X. Wang, Q. Yu, A. Zhang, Enhanced catalytic activity of benzene hydrogenation over nickel confined in carbon nanotubes. *J. Mol. Catal.* 323 (2010) 33–39.

44. A.-G. Boudjahem, W. Bouderbala, M. Bettahar, Benzene hydrogenation over Ni–Cu/SiO$_2$ catalysts prepared by aqueous hydrazine reduction. *Fuel Process. Technol.* 92 (2011) 500–506.

45. Y. Wan, Ch. Chen, W. Xiao, L. Jian, N. Zhang, Ni/MIL-120: An efficient metal–organic framework catalyst for hydrogenation of benzene to cyclohexane. *Micro. Meso. Mater.* 171 (2013) 9–13.

46. R. Wojcieszak, S. Monteverdi, M. Mercy, I. Nowaka, M. Ziolek, M. M. Bettahar, Nickel containing MCM-41 and AlMCM-41 mesoporous molecular sieves: Characteristics and activity in the hydrogenation of benzene. *Appl. Catal.* A 268 (2004) 241–253.

47. Y. Li, L. Zhu, K. Yan, J. Zheng, B. H. Chen, W. Wang, Preparation of Si–Al/α-FeOOH catalyst from an iron-containing waste and surface–catalytic oxidation of methylene blue at neutral pH value in the presence of H$_2$O$_2$. *Chem. Eng. J.* 226 (2013) 166–170.

48. L. Zhu, H. Sun, H. Fu, J. Zheng, N. Zhang, Y. Li, B. H. Chen, Effect of ruthenium nickel bimetallic composition on the catalytic performance for benzene hydrogenation to cyclohexane. *Appl. Catal.* A 499 (2015) 124–132.

49. N. H. H. A. Bakar, M. M. Bettahar, M. A. Bakar, S. Monteverdi, J. Ismail, M. Alnot, PtNi catalysts prepared via borohydride reduction for hydrogenation of benzene. *J. Catal.* 265 (2009) 63–71.

50. C. Qi, F. Jie, L. Wenying, X. Kechang, Synthesis of Ni/Mo/N catalyst and its application in benzene hydrogenation in the presence of thiophene. *Chin. J. Catal.* 34 (2013) 159–166.

51. M. Wang, H. Su, J. Zhou, Sh. Wang, Ru/Al$_2$O$_3$-ZrO$_2$-NiO/cordierite monolithic catalysts for selective hydrogenation of benzene. *Chin. J. Catal.* 34 (2013) 1543–1550.

52. V. H. J. de Beer, J. C. Duchet, R. Prins, The role of cobalt and nickel in hydrodesulfurization: Promoters or catalysts? *J. Catal.* 72 (1981) 369–372.

53. M. Lewandowski, Hydrotreating activity of bulk NiB alloy in model reaction of hydrodenitrogenation of carbazole. *Appl. Catal.* B 168–169 (2015) 322–332.

54. T. C. Ho, Deep HDS of diesel fuel: Chemistry and catalysis. *Catal. Today* 98 (2004) 3–18.

55. L. Peña, D. Valencia, T. Klimova, CoMo/SBA-15 catalysts prepared with EDTA and citric acid and their performance in hydrodesulfurization of dibenzothiophene. *Appl. Catal. B* 147 (2014) 879–887.

56. F. Trejo, M. S. Rana, J. Ancheyta, S. Chávez, Influence of support and supported phases on catalytic functionalities of hydrotreating catalysts. *Fuel* 138 (2014) 104–110.

57. C. F. Linares, G. Acuña, F. Pérez, F. Ocanto, C. Corao, P. Betancourt, J. L. Brito, Garnierite: A new support for hydrodesulfurization catalysts. *Mater. Lett.* 131 (2014) 269–271.

58. Y. C. Park, H.-K. Rhee, The role of nickel in pyridine hydrodenitrogenation over NiMo/Al$_2$O$_3$. *Korean J. Chem. Eng.* 15 (1998) 411–416.

59. E. Rodriguez-Castellon, A. Jimenez-Lopez, D. Eliche-Quesada, Nickel and cobalt promoted tungsten and molybdenum sulfide mesoporous catalysts for hydrodesulfurization. *Fuel* 87 (2008) 1195–1206.

60. M. Martin-Martinez, A. Álvarez-Montero, L. M. Gómez-Sainero, R. T. Baker, J. Palomar, S. Omar, S. Eser, J. J. Rodriguez, Deactivation behavior of Pd/C and Pt/C catalysts in the gas-phase hydrodechlorination of chloromethanes: Structure–reactivity relationship. *Appl. Catal. B* 162 (2015) 532–543.

61. M. Bonarowska, Z. Kaszkur, D. Lomot, M. Rawski, Z. Karpiński, Effect of gold on catalytic behavior of palladium catalysts in hydrodechlorination of tetrachloromethane. *Appl. Catal. B* 162 (2015) 45–56.

62. P. Benito, M. Gregori, S. Andreoli, G. Fornasari, F. Ospitali, S. Millefanti, M. Sol Avila, T. F. Garetto, S. Albonetti, Pd–Cu interaction in Pd/Cu-MCM-41 catalysts: Effect of silica source and metal content. *Catal. Today* 246 (2015) 108–115.

63. S. Lambert, F. Ferauche, A. Brasseur, J.-P. Pirard, B. Heinrichs, Pd–Ag/SiO$_2$ and Pd–Cu/SiO$_2$ cogelled xerogel catalysts for selective hydrodechlorination of 1,2-dichloroethane into ethylene. *Catal. Today* 100 (2005) 283–289.

64. C. Amorim, M. A. Keane, Catalytic hydrodechlorination of chloroaromatic gas streams promoted by Pd and Ni: The role of hydrogen spillover. *J. Hazard. Mater.* 211–212 (2012) 208–217.

65. S. L. Pirard, J. G. Mahy, J.-P. Pirard, B. Heinrichs, L. Raskinet, S. D. Lambert, Development by the sol–gel process of highly dispersed Ni–Cu/SiO$_2$ xerogel catalysts for selective 1,2-dichloroethane hydrodechlorination into ethylene. *Micro. Meso. Mater.* 209 (2015) 197–207.

66. A. Śrębowata, R. Baran, D. Lomot, D. Lisovytskiy, T. Onfroy, S. Dzwigaj, Remarkable effect of postsynthesis preparation procedures on catalytic properties of Ni-loaded BEA zeolites in hydrodechlorination of 1,2-dichloroethane. *Appl. Catal. B* 147 (2014) 208–220.

67. N. Wu, W. Zhang, B. Li, C. Han, Nickel nanoparticles highly dispersed with an ordered distribution in MCM-41 matrix as an efficient catalyst for hydrodechlorination of chlorobenzene. *Micro. Meso. Mater.* 185 (2014) 130–136.

68. A. David, M. A. Vannice, Control of catalytic debenzylation and dehalogenation reactions during liquid-phase reduction by H$_2$. *J. Catal.* 237 (2006) 349–358.

69. F. L. Jia, L. Z. Zhang, X. Y. Shang, Y. Yang, Non-aqueous sol–gel approach towards the controllable synthesis of nickel nanospheres, nanowires, and nanoflowers. *Adv. Mater.* 20 (2008) 1050–1054.

70. L. Bai, F. Yuan, Q. Tang, Synthesis of nickel nanoparticles with uniform size via a modified hydrazine reduction route. *Mater. Lett.* 62 (2008) 2267–2270.

71. Y. Hou, S. Gao, Monodisperse nickel nanoparticles prepared from a monosurfactant system and their magnetic properties. *J. Mater. Chem.* 13 (2003) 1510–1512.

72. D. E. Zhang, X. M. Ni, H. G. Zheng, Y. Li, X. J. Zhang, Z. P. Yang, Synthesis of needle-like nickel nanoparticles in water-in-oil microemulsion. *Mater. Lett.* 59 (2005) 2011–2014.

73. Y. Pan, R. Jia, J. Zhao, J. Liang, Y. Liu, C. Liu, Size-controlled synthesis of monodisperse nickel nanoparticles and investigation of their magnetic and catalytic properties. *Appl. Surf. Sci.* 316 (2014) 276–285.

74. Y. Chen, D. L. Peng, D. Lin, X. Luo, Preparation and magnetic properties of nickel nanoparticles via the thermal decomposition of nickel organometallic precursor in alkylamines. *Nanotechnology* 18 (2007) 505703.

75. N. Stradiotto, K. Toghill, L. Xiao, A. Moshar, R. Compton, The fabrication and characterization of a nickel nanoparticle modified boron doped diamond electrode for electrocatalysis of primary alcohol oxidation. *Electroanalysis* 21 (2009) 2627–2633.

76. G. G. Couto, J. J. Klein, W. H. Schreiner, D. H. Mosca, A.J.A. de Oliveira, A. J. G. Zarbin, Nickel nanoparticles obtained by a modified polyol process: Synthesis, characterization, and magnetic properties. *J. Coll. Interf. Sci.* 311 (2007) 461–468.

77. H. L. Niu, Q. W. Chen, M. Ning, Y. S. Jia, X. J. Wang, Synthesis and one-dimensional self-assembly of acicular nickel nanocrystallites under magnetic fields. *J. Phys. Chem. B* 108 (2004) 3996–3999.

78. Y. Koltypin, A. Fernandez, T. C. Rojas, J. Campora, P. Palma, R. Prozorov, A. Gedanken, Encapsulation of nickel nanoparticles in carbon obtained by the sonochemical decomposition of $Ni(C_8H_{12})_2$. *Chem. Mater.* 11 (1999) 1331–1335.

79. C. Parada, E. Morán, Microwave-assisted synthesis and magnetic study of nanosized Ni/NiO materials. *Chem. Mater.* 18 (2006) 2719–2725.

80. A. Wang, H. Yin, H. Lu, J. Xue, M. Ren, T. Jiang, Effect of organic modifiers on the structure of nickel nanoparticles and catalytic activity in the hydrogenation of *p*-nitrophenol to *p*-aminophenol. *Langmuir* 25 (2009) 12736–12741.

81. Y. Pan, R. Jia, J. Zhao, J. Liang, Y. Liu, C. Liu, Size-controlled synthesis of monodisperse nickel nanoparticles and investigation of their magnetic and catalytic properties. *Appl. Surf. Sci.* 316 (2014) 276–285.

82. J. Xiong, J. Chen, J. Zhang, Liquid-phase hydrogenation of o-chloronitrobenzene over supported nickel catalysts. *Catal. Commun.* 8 (2007) 345–350.

83. R. Baran, I. I. Kamińska, A. Śrębowata, S. Dzwigaj, Selective hydrodechlorination of 1,2-dichloroethane on NiSiBEA zeolite catalyst: Influence of the preparation procedure on a high dispersion of Ni centers. *Micro. Meso. Mater.* 169 (2013) 120–127.

84. A. Śrębowata, R. Baran, D. Lomot, D. Lisovytskiy, T. Onfroy, S. Dzwigaj, Remarkable effect of postsynthesis preparation procedures on catalytic properties of Ni-loaded BEA zeolites in hydrodechlorination of 1,2-dichloroethane. *Appl. Catal. B* 147 (2014) 208–220.

Chapter 3

Hydrogenation by Copper Catalysts

Damian Giziński and Jacinto Sá

Contents

This chapter deals with hydrogenation mediated by copper metal present as supported and unsupported nanoparticles. In the chapter, we highlight synthetic procedures for the preparation of metallic copper nanoparticles and their applications in catalysis. The main target is reactions carried out for fine and pharmaceutical chemicals' production.

3.1 Introduction

During the last decades several highly selective and efficient catalytic methods have emerged for selective and asymmetric hydrogenation, primarily in the field of homogeneous catalysis [1–3]. The majority of the known catalysts are based on transition metal, such as platinum, palladium, rhodium, iridium, and ruthenium. These metals are scarce, expensive, and sometimes significantly toxic. Thus, the search for more economical and environmentally friendly catalysts is an ongoing and still challenging goal. In this respect, novel catalysts based on iron [4–8], copper [9], or zinc [10–12] constitute attractive alternatives.

More specifically, copper is an abundant, relatively nontoxic, and cheap metal mined in around 16 million tons in 2009. Innumerable copper salts and complexes are commercially available or easy to synthesize [9]. In addition, copper is known to be a cofactor in various enzymes and pigments—for example, hemocyanin, which takes part in oxygen transport in living organisms.

While most of the more advanced catalysts containing copper are homogeneous and designed to do asymmetric catalysis, a plethora of heterogeneous copper systems can hydrogenate with high-selectivity organic reactions; some are already 100 years in age. One of those examples is the hydrogenation of benzene by copper catalysts. In 1905, Sabatier and Senderens [13] said that copper could not catalyze the mentioned reaction. This was proven wrong 20 years later by Pease and Purdum [14] and by Ipatieff, Corson, and Kurbatov [15]. The common trend among heterogeneous copper catalysts is that copper needs to be in the metallic state to be active. The ability to prepare and stabilize copper nanoparticles for catalytic applications is an ongoing and still challenging goal, which is reviewed in the subsequent sections of this chapter. Copper nanoparticles are able to perform reactions that, until recently, were unheard of, such as a heterogeneous catalytic "click" cycloaddition [16], which emphasizes its importance and applicability.

3.2 Copper Nanoparticle Preparation

Any particle whose diameter is in a size range of 1 to 100 nm can be called a nanoparticle. As early as the ninth century, to achieve glittering effect artisans used nanomaterials. Since then materials consisting of parts in nanoscale have been applied in various technologies. Their long history might be surprising because, in general, nanoparticles are considered an innovation of modern science. Nevertheless, interest in nanomaterials has risen tremendously in the last decades due to improvement of modern technology. In 1959, American physicist Richard P. Feynman delivered a lecture titled "Plenty of Room on the Bottom." It was a prediction about nanotechnology. Because of their small size, nanoparticles exhibit unique properties. Not only size determines properties and applications. As it will be shown later, many aspects have to be considered. There are plenty of methods to obtain materials with different properties, such as the magnetic or crystalline structure. An immense number of preparation ways allows looking for new applications of nanoparticles and provides a lot of opportunities.

Every pathway of metal nanoparticle synthesis boils down to reducing metal ions to stable zero-valent metal. Over the past few years, many manners of reduction have been extensively investigated. It has been proven that all thermal processes, sonification, microwave irradiation, or chemical processes can be successfully applied in nanocopper synthesis. In general, a couple of methods are used simultaneously. For instance, chemical reduction can be more effective in higher temperatures or under irradiation conditions.

In 1857, Faraday published his work describing how to achieve zero-valent metal colloids. It relies on chemical reduction of transition metal with a stabilizing agent. Since then it

has become one of the most commonly used methods to prepare nanoparticles. In itself reduction is a first step. Obviously,
at this stage, metal atoms are arising. Nonetheless, in the
reactor there are still metal ions, so both ions and zero-valent
atoms are present in solution. They can collide with each
other and, as a result, a stable nucleus appears; this is known
as a nucleation process. Because of the collision with metal
ions, atoms, or different nuclei, nuclei may grow up, which is
a final step of producing nanoparticles.

Faraday expressed that some kind of protective agent has
to be applied to stabilize structure and to prevent agglomeration. These agents influence nanoparticles in two basic ways
[17] (Figure 3.1). In the first case nanoparticles get covered by
ions and counter-ions, which form an electric double layer.
Covered particles repulse each other via coulombic forces.
Therefore, this is an electrostatic stabilization. In the second
case nanoparticles get covered by sterically demanding organic
molecules. This causes specific protection of the nanoparticles'
surface and it prevents agglomeration by steric effect. This
phenomenon is called steric stabilization.

As mentioned previously, the method consisting of reducing
a copper salt by an electron donor agent is the most widely

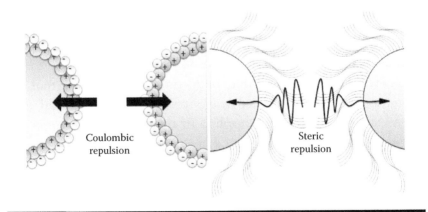

**Figure 3.1   Two main modes of stabilization. Electrostatic stabiliza-
tion on the left and stabilization by steric repulsion on the right.**

used method, and NaBH$_4$ (sodium borohydride) is one of the most popular reducing agents [18]. Addition of BH$_4^-$ ions causes immediate reduction, which can be monitored by following the changes in solution color. Presence of metallic copper can be easily proved by an optical technique such as UV/Vis spectroscopy. The plasmon resonance of copper nanoparticles appears at ~560 nm (Figure 3.2).

In the Dang approach [19] for fine copper nanoparticles' preparation, ascorbic acid was used as an antioxidant agent. Reoxidation is one of the factors that disturb the whole process. Ascorbic acid is well known for its antioxidation

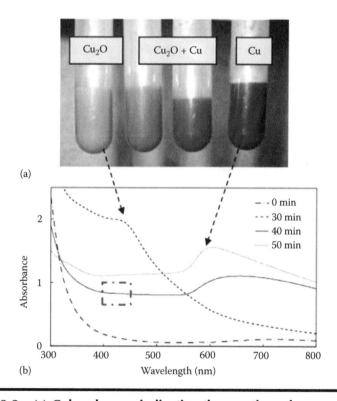

Figure 3.2 (a) Color changes indicating the reaction advancement; (b) UV–Vis absorbance spectra corresponding to different reaction stages; surface plasmon resonance bands of Cu^0 and of Cu$_2$O are indicated by the arrows. (Reproduced from M. Blosi, S. Albonetti, M. Dondi, C. Martelli, G. Baldi, *J. Nanoparticle Res.* 13 (2010) 127–138. With permission.)

properties that work as oxygen molecule scavengers, a strategy often applied in nanoparticle synthesis [19–21]. Chemical antioxidants are not the only way to prevent oxidation. A simple manner is to prevent oxidation in the first place by changing experimental conditions—for example, carrying out the synthesis under an anaerobic atmosphere instead of ambient conditions [22–24]. Gutiérrez et al. [24] have shown a synthesis pathway in a nitrogen atmosphere. In this approach lithium sand was used as a reducing agent. To keep anaerobic conditions, the whole process was carried out in a Parr reactor. This procedure led to obtaining Cu nanoparticles without any significant reoxidation occurring.

Plenty of aspects and conditions can be improved or changed in the reduction methodology to attain a simple procedure that is in accordance with green chemistry postulates. One of these modifications resulted in a specific kind of chemical reduction method: polyols reduction. This considerable group of compounds includes alcohols containing multiple hydroxyl groups, which are useful to nanoparticle synthesis due to their chelating and reducing abilities [25]. In 2004, Zhao with co-workers [26], presented their route to prepare copper nanocrystals in which they used ethylene glycol both as a solvent and stabilizer. The suspension resulting from the mixture of copper salt and polyol was microwaved under ambient conditions for 15 minutes. According to x-ray diffraction (XRD) measurements, all CuO, Cu_2O, CuO/Cu_2O, Cu_2O/Cu, and Cu formations were detected. From TEM (transmission electron microscopy) images, the pure copper nanocrystals had sizes ranging from 100 to 120 nm. This simple and environmentally friendly method showed methodology potential, but from a catalytic point of view, the uncompleted reduction and relatively large nanoparticle size prevented the exploitation to perform hydrogenations.

A few years later Blosi et al. [25] amended this approach. They used polyol (dietylene glycol) as a reductant and PVP E30 (polyvinylpyrrolidone) as a stabilization agent, which they heated to the desired reaction temperature. Only then did they add the copper salt and antioxidant agent. The Blosi method enabled them to prepare copper nanoparticles ~40 nm in size (Figure 3.3). However, the final nanoparticles' morphology was very sensitive to reaction conditions (e.g., reagent addition temperature). For example, longer heating ramp nucleation lasts longer and bigger nanoparticles are formed, while a shorter heating ramp or addition of a

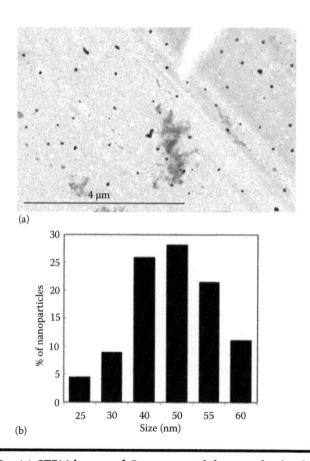

(a)

(b)

Figure 3.3 (a) STEM image of Cu nanoparticles, synthesized with $[Cu^{2+}] = 16.6$ mM, $[nPVP/nCu^{2+}] = 5$, $[nC_6H_8O_6/nCu(ac)_2] = 2.5$, T = 100°C, t = 10 min; (b) particle size distribution calculated by STEM analysis, average diameter: 46 nm, standard deviation: 9 nm. (Reproduced from M. Blosi, S. Albonetti, M. Dondi, C. Martelli, G. Baldi, *J. Nanoparticle Res.* 13 (2010) 127–138. With permission.)

copper source in the destined temperature resulted in smaller nanoparticles (Figure 3.4).

Kawasaki et al. [27] showed a quite different approach, which was to enable preparation of copper nanocrystals with 3 nm. In contrast to previous methods, their synthesis was carried out in nonaqueous polyol solution with NaOH under nitrogen atmosphere without using any additional protective agent. Through all these antioxidation operations and reaction temperature at 185°C, small copper nanoparticles were obtained via a microwave-assisted polyol method. With reference to x-ray adsorption spectroscopy measurements, no adsorption peak of Cu-O stretches was found, indicating complete reduction of a sample. The authors proposed an explanation of how Cu nanocrystals were formed using their method. MALDI-MS results confirmed that the nanoparticles' surfaces were covered by polyethylene glycol, which was responsible for agglomeration prevention.

Dhas, Raj, and Gedanken [23] compared two methods to prepare Cu nanoparticles: thermal and sonification. In both cases a copper hydrazine carboxylate precursor was used. In the thermal process, the solution was stirred for 3 h at 80°C in argon atmosphere. They observed a change in solution color from blue to red suggestive of metallic copper formation. In the sonication process, the solution was exposed to high-intensity ultrasound radiation. According to XRD analysis, the sonochemical process pathway gave smaller Cu nanoparticles.

As has been reported [28–31], well-prepared copper nanoparticles may be prepared from a thermal decomposition method. In general, this method consists of reaction between metal salt and a capping agent that makes a copper-containing complex. In the next step, when the complex reaches its decomposition temperature, nanoparticles are created as one of the products. Kim et al. [28] described their path of Cu nanoparticle production by a thermal decomposition method. In this procedure Cu-oleate complex was formed as a result of a $CuCl_2$ and sodium oleate reaction. Next, the

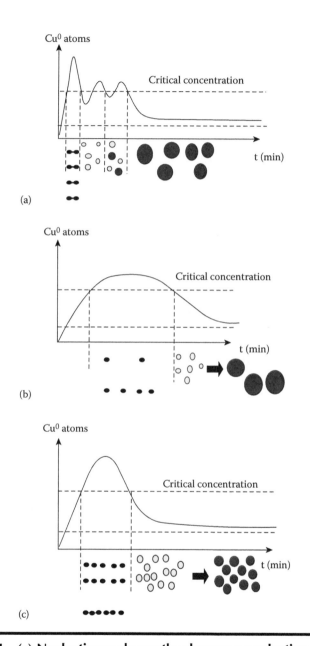

Figure 3.4 (a) Nucleation and growth schemes: a nucleation speed too high, multistep nucleation; (b) nucleation speed too low; (c) optimized experimental conditions: nucleation and growth equilibrium. (Reproduced from M. Blosi, S. Albonetti, M. Dondi, C. Martelli, G. Baldi, *J. Nanoparticle Res.* 13 (2010) 127–138. With permission.)

complex was put into a container with high heat resistance and the temperature was raised to 290°C. It turned out that nanoparticle size depended on sodium oleate concentration (i.e., smaller particles were obtained when lower amounts of capping agent were used). With reference to TEM images, the smallest particles obtained had diameters around 13 nm. In 2014, Betancourt-Galindo et al. [32] showed how to modify the sodium oleate method to attain even smaller nanoparticles. In the Betancourt-Galindo approach, phenyl ether was used. It was noticed that spherical copper nanoparticles in size ranges of 4 to 18 nm were evidenced by XRD and TEM measurements.

Another widely used pathway for obtaining copper nanoparticles is based on an electrolysis process. An electrochemical route as a simple way of producing nanocopper has been reported. Figure 3.5 shows the overall procedure of this path. A typical experiment consists of two electrode processes: oxidative dissolution on an anode and metal ion reduction on a cathode. In a future period as well as in previous methods,

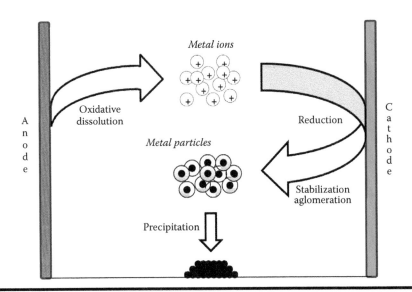

Figure 3.5 Schematic representation of electrochemical methodology for the preparation of metallic copper nanoparticles.

nanoparticles are being subjected to agglomeration prevention and stabilization. Hashemipour et al. [33] presented a default way of creating electrochemical Cu nanoparticles. It was shown that cathodic potential influenced reduction velocity and, consequently, nanoparticle growth.

It should be mentioned that copper nanoparticles need stabilization and protection against sintering and oxidation, respectively. This can be achieved by supporting them as in classic heterogeneous catalysis or by surrounding them with polymeric resins. In both cases apart of stabilization and protection, the supports confer the nominal size required for continuous flow hydrogenation reactions.

3.3 Heterogeneous Hydrogenation Reactions

3.3.1 Hydrogenation of Cinnamaldehyde

Copper-based catalysts are very efficient for many catalytic reactions, including selective hydrogenation, oxidation, and amination. A desirable catalyst should, therefore, be selective to hydrogenate the C=O bond and indicate the ability to avoid formation of unwanted by-products. However, noble metals hydrogenate the C=C bond faster than the C=O group [34,35]. This can be changed by addition of promoters, which might improve the selectivity to unsaturated alcohols (e.g., in hydrogenation of α,β-unsaturated aldehydes). These products are widely used as flavors and perfumes, as well as intermediates for the synthesis of valuable organic molecules.

Selective hydrogenation of cinnamaldehyde to the corresponding alcohols is of great importance in the fine chemicals industry. Also, it is considered a good example of reaction in which the catalytic behavior is correlated with microstructures of heterogeneous catalysts. The reaction is also interesting from a fundamental point of view because it poses a problem of region- and chemoselectivity [36]. Generally, cinnamaldehyde (CAL) is hydrogenated to cinnamyl alcohol (COL) and hydrocinnamaldehyde (SAL), depending on whether the C=C or C=O double bond is hydrogenated. Subsequently, hydrocinnamaldehyde and cinnamyl alcohol can both be hydrogenated to hydrocinnamyl alcohol (SOL) as the final product. Figure 3.6 shows possible pathways for cinnamaldehyde hydrogenation.

Marchi et al. [36] performed cinnamaldehyde hydrogenation with various Cu-based catalysts. Copper supported on silica was used as a basic one. The authors also prepared a series of materials modified by additions of different metals. It has been shown that the metal active phase and support influence catalyst selectivity (Figure 3.7). In reference to this work, monometallic copper on silica caused C=C hydrogenation to predominate over the reduction of the C=O bond (79.6% selectivity to hydrocinnamaldehyde).

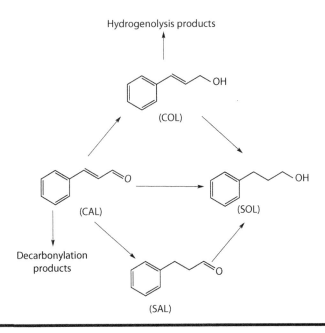

Figure 3.6 Cinnamaldehyde hydrogenation reactions. (Reproduced from A. J. Marchi, D. A. Gordo, A. F. Trasarti, C. R. Apesteguia, *Appl. Catal. A* 249 (2003) 53–67. With permission.)

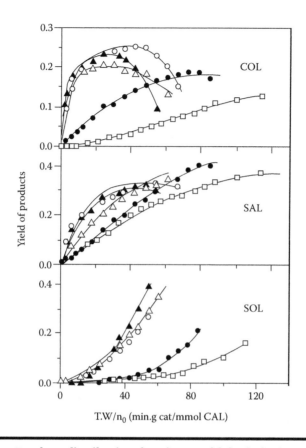

Figure 3.7 **Product distribution for cinnamaldehyde hydrogenation reactions on Cu-based catalysts. Product yield (η_i) as a function of parameter Wt/n_T^0 [393 K, 10 bar, 1 g catalyst]: Cu/SiO$_2$ (□), Cu–Al (●), Cu–Zn–Al (○), Cu–Ni–Zn–Al (▲), Cu–Co–Zn–Al (△). (Reproduced from A. J. Marchi, D. A. Gordo, A. F. Trasarti, C. R. Apesteguia, *Appl. Catal. A* 249 (2003) 53–67. With permission.)**

It has been adopted that cinnamaldehyde adsorption over copper occurs via π_{CC} and/or d_i–π adsorption modes. Additionally, both of them are influenced by the repulsive forces existing between copper d-orbitals and the phenyl group of cinnamaldehyde (Figure 3.8a). At the beginning, cinnamaldehyde is selectively hydrogenated to hydrocinnamaldehyde and then, after an induction period, formation of cinnamyl alcohol is detected. Inhibition of the C=O group hydrogenation, in the initial period,

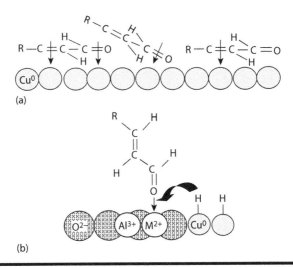

(a)

(b)

Figure 3.8 Cinnamaldehyde adsorption over Cu-based catalysts: (a) over pure copper and (b) over copper doped by various ions. (Reproduced from A. J. Marchi, D. A. Gordo, A. F. Trasarti, C. R. Apesteguia, *Appl. Catal. A* 249 (2003) 53–67. With permission.)

seems to be caused by copper surface saturation. Next, clean metal surface patches allow the cinnamaldehyde molecules to interact with the copper surface via $d_i-\pi_{CO}$ or π_{CO} adsorption modes and selectivity to cinnamyl alcohol rising (Figure 3.8a). Nevertheless, the hydrocinnamaldehyde formation rate remains significantly higher than that of cinnamyl alcohol during the entire catalytic test.

As previously mentioned, well-prepared catalysts should exhibit high selectivity to unsaturated alcohols (as valuable reagents in the chemical industry) and prevent formation of unwanted by-products. Liu at al. [37] prepared copper-containing, ordered mesoporous carbons (Cu-OMCs) via a one-pot aqueous self-assembly soft templating route followed by a direct carbonization process for selective hydrogenation of cinnamaldehyde (Figure 3.9). In this work, three hydrogenation products were detected: cinnamyl alcohol, hydrocinnamaldehyde, and hydrocinnamyl alcohol. Moreover, some unsolicited products were also observed. For example, the presence of a negligible amount

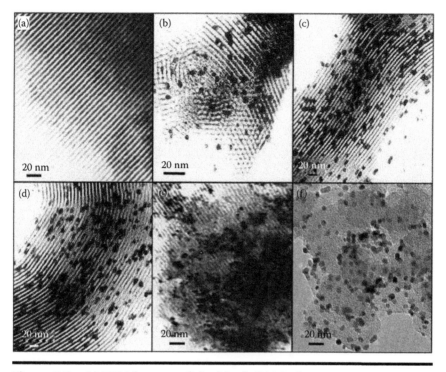

Figure 3.9 (a) TEM images of the blank OMC; (b) 4.5 Cu-OMC; (c) 7.2 Cu-OMC; (d) 9.6 Cu-OMC; (e) 10.2 Cu/OMC; (f) 10.4 Cu/AC. (Reproduced from Z. Liu, Y. Yang, J. Mi, X. Tan, Y. Song, *Catal. Commun.* **21** (2012) 58–62. With permission.)

of ethylbenzene showed that the decarbonylation of cinnamaldehyde had occurred. Therefore, in this case, formation of unwanted products slightly decreases the applicability of these catalysts, in terms of previous assumptions. On the other hand, using these materials resulted in high selective hydrogenation of cinnamaldehyde (Figure 3.10). It was found that the main product over the 9.6 Cu-OMC catalyst was cinnamyl alcohol arising from the C=O hydrogenation. When the 10.2 Cu/OMC was used as the catalyst, the main product was still cinnamyl alcohol, but the selectivities to hydrocinnamaldehyde and hydrocinnamyl alcohol obviously increased (>5%). In addition, the product selectivities over the 9.6 Cu-OMC and 10.2 Cu/OMC catalysts remained fairly constant with increasing total conversion of cinnamaldehyde.

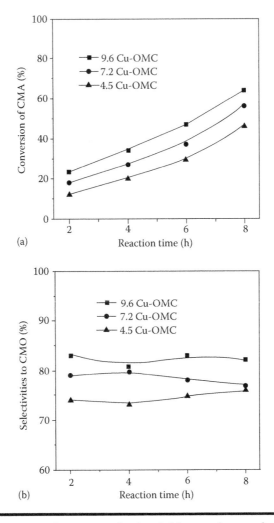

(a)

(b)

Figure 3.10 Conversions (a) and selectivities to cinnamyl alcohol (b) of hydrogenation of cinnamaldehyde over 4.5 Cu-OMC, 7.2 Cu-OMC, and 9.6 Cu-OMC catalysts. (Reproduced from Z. Liu, Y. Yang, J. Mi, X. Tan, Y. Song, *Catal. Commun.* **21** (2012) 58–62. With permission.)

For comparison, commercially activated carbon (marked as AC; micropores pose 80% of the whole surface) was used as support and then checked in the same reaction. It was found that when 10.7 Cu/AC was used, selectivity to cinnamyl alcohol reached a level as high as 60%. The microporous structure was the limiting agent to the diffusion of an intermediate product of cinnamyl alcohol from the catalyst, resulting in longer contact time and making hydrogenation to saturated alcohol more successful. Such a high selectivity was also attributed to both a good dispersion of the copper particles on the supports and a strong interaction between Cu and the support [38,39].

For OMC supported catalysts, high activity was attributed to the intimate interfacial contact between the active particles and the carbon support. Highly dispersed Cu nanoparticles formed with the organic–organic self-assembly strategy (OSS) create strong interaction with the carbon support. Furthermore, the ordered mesoporous structure of the OMC made mass transfer easier compared to the commercial AC and provided better accessibility to reactant molecules. The multifunction of the Cu-OMC catalyst synthesized with OSS should be responsible for the superior performance of selective hydrogenation of cinnamaldehyde.

Previous examples have reported the origin of copper/support catalyst selectivity in hydrogenation of cinnamaldehyde over supported and unsupported Cu nanoparticles [24]. Cu nanoparticles are highly selective (Figure 3.11) for the hydrogenation of the C=O bond when the C=C bond is present in the

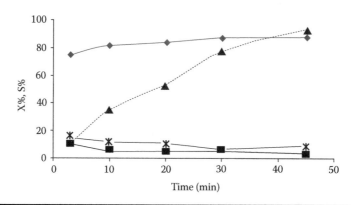

Figure 3.11 **Dependence on time of conversion (▲) of cinnamaldehyde and of the selectivities to cinnamyl alcohol (◆), hydrocinnamaldehyde (✗) and hydrocinnamyl alcohol (■) for Cu nanoparticles. (Reproduced from V. Gutiérrez, F. Nador, G. Radivoy, M. A. Volpe, *Appl. Catal. A* 464–465 (2013) 109–115. With permission.)**

same molecule. In addition, during reaction time, the selectivities corresponding to the different primary hydrogenation products were stable on constant levels. No unwanted byproducts of decarbonilation or polymerization were detected. Such behavior (according to previous assumptions about high selectivity of C=O hydrogenation and absence of undesired products) was ascribed to both the nanosized dimension and absence of support.

It is worth mentioning that the main problems of unsupported Cu nanoparticles were separation and deactivation. The latter relates to the oxidation of the active sites. Reutilization is one of most important advantages of heterogeneous catalysis and should be considered from the catalyst preparation stage. Thus, Gutiérrez et al. [24] supported copper nanoparticles on mesoporous MCM-41, SiO_2, and CeO_2. Figures 3.12 and 3.13 show the catalytic behavior for some of the supported catalysts. Unfortunately, the supported systems did not achieve selectivities similar to those of the unsupported, especially in the case of SiO_2 (Figure 3.13). It should be

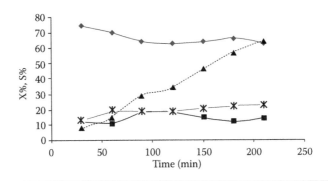

Figure 3.12 Dependence on time of conversion (▲) of cinnamaldehyde and of the selectivities to cinnamyl alcohol (◆), hydrocinnamaldehyde (✱) and hydrocinnamyl alcohol (■) for Cu nanoparticles/ MCM-4. (Reproduced from V. Gutiérrez, F. Nador, G. Radivoy, M. A. Volpe, *Appl. Catal. A* 464–465 (2013) 109–115. With permission.)

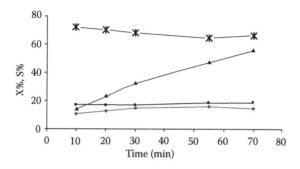

Figure 3.13 **Dependence on time of conversion (▲) of cinnamalde-hyde and of the selectivities to cinnamyl alcohol (◆), hydrocinnamal-dehyde (✻) and hydrocinnamyl alcohol (■) for Cu nanopartices/SiO₂. (Reproduced from V. Gutiérrez, F. Nador, G. Radivoy, M. A. Volpe,** *Appl. Catal. A* **464–465 (2013) 109–115. With permission.)**

mentioned that TEM images showed that copper nominal size was similar for supported and unsupported catalysts.

Copper seems to comply with suitable catalyst requirements in hydrogenation of α,β-unsaturated aldehydes. Well-prepared Cu nanoparticles exhibit high selectivity to hydrogenated C=O bonds. It should not be omitted that, for industry, cinnamaldehyde hydrogenation catalysts need to be active, very selective, and reusable. Specific supports are required to attain industrial requirements.

3.3.2 Hydrogenation of Levulinic Acid

Levulinic acid (LA) is a well-known product due to its great importance in the new generation of biofuels. It can be produced by acid hydrolysis of both lignocellulosic materials and urban residues [40]. Good yields of levulinic acid can be achieved not only in laboratory scale but also in semi-industrial scale. An environmentally friendly manufacturing procedure is advantageous, plus it can produce valuable products from further processing (e.g., γ-valerolactone [GVL]).

Gamma-valerolactone possesses several properties that make it a sustainable liquid that could be used for both energy and carbon-based product manufacturing [41]. Moreover, γ-valerolactone exhibits almost the same behavior as ethanol when added to gasoline. Thus, there is increased interest in improvement of γ-valerolactone production. According to the literature, selective hydrogenation over a specific catalyst seems to be one of the easiest ways. Mainly, hydrogenation of levulinic acid to γ-valerolactone has been carried out over ruthenium-based catalysts [42–44]. Nevertheless, because of availability and lower cost, non-noble metals were also investigated in this reaction. As expected, there are occasionally reports of using non-noble metals for levulinic acid hydrogenation.

Upare et al. [45] performed vapor-phase hydrogenation of levulinic acid with a copper-silica catalyst. In their approach, formic acid acted as a hydrogen source. A small amount of copper (ca. 10 wt%) was sufficient for conversion of 66% of levulinic acid into γ-valerolactone (45%) and angelica-lactone (55%). To increase selectivity, all copper/silica weight ratios, reaction temperature, and levulinic acid/formic acid molar ratios were checked. The reaction was carried out at quite a high temperature (265°C); however, it was performed under 1 atmosphere of nitrogen. To improve conversion, higher loadings of Cu on silica (80 wt%) were needed. These catalysts produced 81% γ-valerolactone and 19% angelica-lactone, with much better conversion levels.

Also, in the work of Patrakumar et al. [46], catalysts with different amounts of copper were investigated in levulinic acid hydrogenation. Gamma-Al_2O_3 was used as a support and copper loading was varied from 2 to 20 wt%. The rationale to use γ-Al_2O_3 relates to its acidity, both Lewis and Bronsted sites, which carries out the dehydration reaction. This is the first step of levulinic hydrogenation to γ-valerolactone (Figure 3.14). The sample with the lowest metal loading converted 82% of initial levulinic acid, which increased even further to around

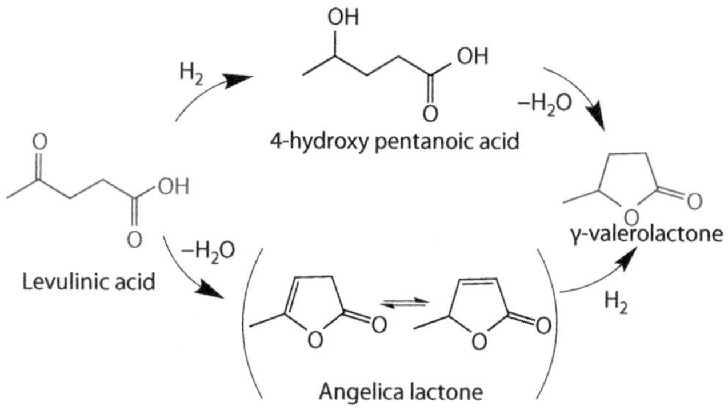

Figure 3.14 Reaction pathway of hydrogenation of levulinic acid to γ-valerolactone.

98% when copper loading was 5 wt%. Further increase in copper loading from 10 to 20 wt% decreased the conversion of levulinic acid to 41%. The decrease in conversion of the catalysts with increase in copper loading was attributed to the lack of availability of acidic sites.

Also, the best selectivity to γ-valerolactone was 5 wt% Cu on alumina. As with the conversion, the selectivity of γ-valerolactone increased from 76% to 87% with increase in copper loading from 2 to 5 wt%; however, it decreased from 87% to 56% at higher Cu loadings. This behavior indicates that the selectivity of γ-valerolactone is directly related to copper dispersion on Al_2O_3. Similarly to the previous example, optimal reaction temperature has been established at 265°C. Also, Obregón and co-workers [47] have noticed good properties of combining copper with alumina for γ-valerolactone producing. Their work, has investigated a series of both monometallic and bimetallic catalysts. According to thermogravimetry (TGA) measurements, another advantage of levulinic acid hydrogenation over copper was found. By comparing TGA curves of fresh and used catalysts, it can be clearly seen that the peak of carbon decompositions was smaller for copper-based catalysts. This implies that copper plays a significant role in preventing catalysts' surfaces from carbon decompositions.

3.3.3 Other Hydrogenations over Cu Catalysts

3.3.3.1 Hydrogenation of Furfural

There are several examples corroborating that copper is a suitable metal for other selective hydrogenation reactions than the ones presented so far. Here, we show a brief overview of other hydrogenation processes that employ copper catalysts. As a first instance, hydrogenation of furfural will be quoted. Mainly, in industry scale, a copper-chromium catalyst has been used in this reaction. The greatest disadvantage of these materials is their toxicity and damaging influence on the environment. As has been shown [48–52], the use of copper instead of Cr-based materials leads to even 100% selectivity to the desired product—namely, furfuryl alcohol. Furfuryl alcohol is widely used for the production of resins (liquid and thermostatic), various synthetic fibers, rubber resins, and farm chemicals. It is also a required chemical intermediate for the production of lysine, vitamin C, lubricants, dispersing agents, and plasticizers.

In 2003, Nagaraja et al. [48] investigated a Cu-MgO catalyst in the selective hydrogenation of furfural to furfuryl alcohol (Figure 3.15). It should be mentioned that the reaction could be carried out in vapor and liquid phase hydrogenation. In the Nagaraja group's contribution, the vapor process was reported. They used catalysts prepared in a different manner—namely, via coprecipitation, impregnation, and solid–solid wetting methods. Catalysts made via the first method exhibited high metal dispersion, which led to high conversions of furfural (98%) with high selectivity toward furfuryl alcohol (98%).

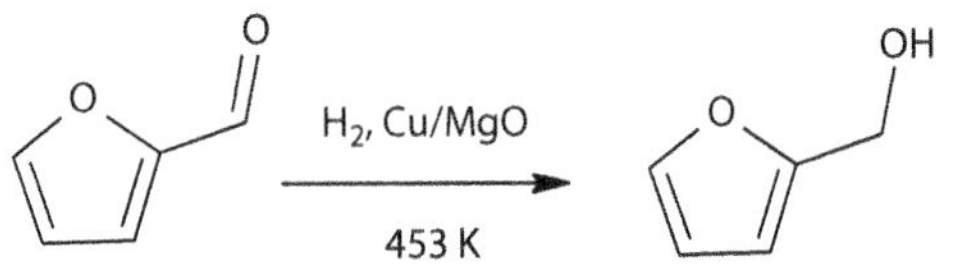

Figure 3.15 **Reaction scheme of furfural hydrogenation over copper supported by magnesium oxide.**

Moreover, during the 5 h reaction the authors saw no significant deactivation.

A few years later the same group published work [49] reporting the effect of Cu loading and reaction temperature on furfural hydrogenation catalyzed by Cu/MgO. Generally, selectivity to furfural alcohol might be assigned to

- Interactions between metal and support
- Electronic and steric influence of the support
- Morphology of the metal particles

Other factors such as steric effects of substituents at the conjugated double bond, effect of pressure, and selective poisoning may also contribute. Selective hydrogenation of the C=O bond while keeping the C=C bond in the furan ring unsaturated is clearly a crucial advantage of copper-based catalysts. In this particular case, the authors' explanation for the high selectivity of coprecipitated catalysts to furfural alcohol relates to the presence of Cu^0 metal species and the oxygen vacancies of MgO. It was also found that Cu loading affected a catalyst's behavior only in terms of conversion. The highest furfural conversions were obtained for hydrogenation over 16 wt% Cu/MgO catalysts. From a selectivity point, 453 K was the most suitable reaction temperature. Higher temperature activates the C=C bond and a bigger amount of undesired products could be obtained. Recently, it was shown that SBA-15 [50] is a good catalyst support for the selective manufacturing of furfural alcohol in the vapor phase, in particular when a copper loading of 15 wt% is used.

As Villaverde et al. [51] have demonstrated, copper catalysts can be applied in a liquid phase as well as in a vapor phase. Two types of copper-based catalysts have been investigated: copper supported by commercially available silica (Cu/SiO$_2$) and Cu-Zn-Al and Cu-Mg-Al catalysts prepared by the coprecipitation method. For comparison, Cu-Cr material was synthesized. It was found that Cu-Mg-Al was the best catalyst of the series in selective hydrogenation of furfural to furfuryl alcohol. The authors suggested that the carbonyl group of furfural adsorbs over Lewis acid sites (Mg^{2+}, Al^{3+}) on the surface similar to spinel (magnesium and aluminum oxide mineral). Thus, the atomic composition of this surface has an impact on its adsorptive properties. In other words, the nature of cations in the spinel plays a very important role.

The Cu-Zn-Al and Cu-Mg-Al have a spinel-like phase that determines the metal electronic density and surface crystallographic planes. The effect of interaction between the copper catalytic activity and surface seems to be more important when Mg^{2+} ions are present on this surface. This also gives rise to surface catalytic activity for furfural chemisorption and the subsequent hydrogenation. Additionally, both the small crystallite size (~5 nm) and low crystallinity degree for catalysts formed during the coprecipitation method (with Mg and Zn) empowered the mentioned interaction. It has to be said that all catalysts used in these tests showed 100% selectivity to alcohol. In further work [52], the same group, after selecting the most suitable catalyst, determined optimal conditions for furfural hydrogenation in the liquid phase.

Not only was the impact of temperature and copper loading checked but also the outcome of using different sources of hydrogen. According to the results, totally selective and effective hydrogenation of furfural was possible in 423 K over Cu-Mg-Al catalysts with 40 wt% of copper; 2-propanol acted as the most productive H_2 feeder of all those that were considered.

3.3.3.2 Hydrogenation of Dimethyl Oxalate

Ethylene glycol, the main product of dimethyl oxalate hydrogenation, is a very important chemical intermediate [53], which is widely used in polyester, dynamite, and antifreeze manufacturing. Copper, because of high hydrogenation activity of esters to alcohols, has been used in this process also. Basically, copper supported on SiO_2 shows good performance in this case, although other materials like ZrO, Al_2O_3 [54], hexagonal mesoporous silica [55], and, very recently, hydroxyapatite [56] have also been applied. All these catalysts can be prepared via different methods; nevertheless, for this particular application, only few of them have been employed. Chen et al. [57] demonstrated dimethyl oxalate hydrogenation (Figure 3.16) over Cu/SiO_2 catalysts obtained by an ammonia evaporation method. It was found that preparation temperature had a bearing on catalysts' evolution [57,58].

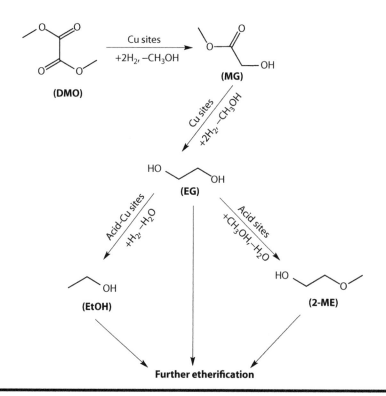

Figure 3.16 **Reaction pathways for the conversion of dimethyl oxalate over the copper catalysts in H_2 flow. (Reproduced from Y. Zhu, Y. Zhu, G. Ding, S. Zhu, S. Zheng, Y. Li, *Appl. Catal. A* 468 (2013) 296–304. With permission.)**

It was ascertained that Cu ions are responsible for selective hydrogenation of the C=O bond, because it may influence an oxygen's lone electron pair and thus change C=O polarization. Tables 3.1 to 3.3 summarize the physico-chemical parameters and catalytic performance of the synthesized of the catalysts. Both dispersion of copper species and ratio of Cu^0/Cu^+ have impact on the catalytic performance. Therefore, the catalyst prepared at 363 K, which had the highest $Cu^+/Cu^+ + Cu^0$ and the largest Cu^+ surface area, allowed highest hydrogenation activity. Over this catalyst, 98% selectivity to ethylene glycol with total conversion was achieved [57].

For comparison, results of hydrogenation over copper-silica catalysts made via the urea hydrolysis method [31] will be shown. It is worth mentioning that the preparation method required a relatively low temperature for catalyst reduction. Namely, the first experiment was carried out with a catalyst reduced at 350°C and then reduction temperature was decreased to 200°C. It turned out that the catalyst reduced in milder conditions was able to convert 100% of dimethyl

Table 3.1 Physicochemical Properties of the Reduced Cu/SiO$_2$ Samples

Catalyst	SBET (m^2g^{-1})	Vp (cm^3g^{-1})	dp (nm)	dCu^a (nm)	dCu^c (nm)
CuSi-333	132	0.50	14.0	22.1, 5.2[b]	30.2, 6.4
CuSi-343	181	0.75	13.5	15.7, 4.8[b]	14.3, 6.1
CuSi-353	248	0.77	12.3	4.1	7.1
CuSi-363	281	0.71	9.0	3.6	7.3
CuSi-373	309	0.96	10.3	3.9	7.2

Source: Reproduced from L. Chen, P. Guo, M. Qiao, S. Yan, H. Li, W. Shen, H. Xu, K. Han, *J. Catal.* 257 (2008) 172–180. With permission.

[a] CuO crystallite size calculated by the Scherrer formula.
[b] Bimodal size distribution of Cu crystallites calculated by fitting the (111) diffraction peak of fcc Cu.
[c] Cu particle size measured by TEM.

Table 3.2 Cu Species on the Reduced Cu/SiO$_2$ Catalyst Derived from Cu LMM XAES Spectra

Catalyst	KE (eV)		α′ (eV)		XCu^{+a} (%)	SCu^{0b} (m^{2}g^{-1})	SCu^{+c} (m^{2}g^{-1})
	Cu$^+$	Cu^0	Cu$^+$	Cu^0			
CuSi-333	914.4	918.4	1847.0	1851.0	40.4	3.6	2.4
CuSi-343	914.2	918.2	1847.0	1851.0	40.5	7.8	5.3
CuSi-353	914.2	918.0	1846.9	1850.7	45.7	9.6	8.1
CuSi-363	914.3	918.4	1846.9	1851.0	54.9	9.2	11.2
CuSi-373	914.2	918.0	1846.8	1850.6	52.0	9.8	10.6

Source: Reproduced from L. Chen, P. Guo, M. Qiao, S. Yan, H. Li, W. Shen, H. Xu, K. Han, *J. Catal.* 257 (2008) 172–180. With permission.

[a] Intensity ratio between Cu$^+$ and (Cu$^+$ + Cu^0) by deconvolution of Cu LMM XAES spectra.
[b] Metallic Cu surface area determined by N$_2$O titration.
[c] Calculation based on XCu$^+$ and SCu^0 assuming that Cu$^+$ ion occupies the same area as that of the Cu^0 atom and has the same atomic sensitivity factor as that of Cu^0.

Table 3.3 Catalytic Performance of Cu/SiO$_2$ Catalysts Prepared by the Ammonia Evaporation Method in Gas-Phase Hydrogenation of Dimethyl Oxalate

Catalyst	DMO conversion (%)	Selectivity (%)			
		MG	EG	Ethanol	1,2-BDO
CuSi-333	34	82	18	0.7	0.2
CuSi-343	46	73	26	1.5	0
CuSi-353	65	65	33	1.7	0
CuSi-363	79	57	43	1.8	0
CuSi-373	73	74	25	1.1	0.1

Source: Reproduced from L. Chen, P. Guo, M. Qiao, S. Yan, H. Li, W. Shen, H. Xu, K. Han, *J. Catal.* 257 (2008) 172–180. With permission.

Note: Selectivity to methyl glycolate (MG), ethylene glycol (EG), ethanol, and 1,2-butanediol (1,2-BDO).

[a] Reaction conditions: p = 2.5 MPa, T = 473 K, H$_2$/DMO = 50 (mol mol^{-1}), and LHSV of DMO = 0.50 h^{-1}.

oxalate. That may have significant meaning for industry in the context of catalyst stabilization. Additionally, complete conversion was achieved only for proper copper loading (~20%). The urea hydrolysis method is also an effective way to obtain ethylene glycol. The selectivity of this product reached 90% over 20% Cu/SiO_2 catalyst.

A copper catalyst could also be modified for changing its selectivity. Zhao et al. [59] reported a pathway for ethanol synthesis via selective hydrogenation of dimethyl oxalate. This work effect of boron addition was also investigated. Copper-based catalysts were prepared by an ammonia-evaporation method and boron modification was obtained by impregnation. All catalysts achieved 100% dimethyl oxalate conversion. Nonetheless, selectivity to ethanol changed along the boron loading. It increased up to the catalyst with Cu/B = 3 (85.6%) and decreased for higher Cu/B molar ratios (Figure 3.17). Catalysts with a suitable amount of boric oxide have significantly improved dispersion of copper species. After

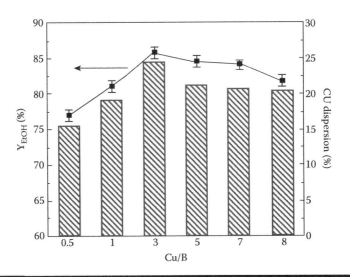

Figure 3.17   Ethanol (EtOH) yield and Cu dispersion of catalysts with different Cu/B. (Reproduced from S. Zhao, H. Yue, Y. Zhao, B. Wang, Y. Geng, J. Lv, S. Wang, J. Gong, X. Ma, *J. Catal.* 297 (2013) 142–150. With permission.)

reaching the maximum, high boron loading caused a decline in the copper surface area. High copper dispersion and a large metallic copper surface seem to be crucial for producing chemo-selective ethanol.

To see stabilization changes with boron doping, Cu/SiO_2 and CuB/SiO_2 were treated at high temperatures (723 K) for 12 h. It is known that, because of weak interaction between copper and support, some unwanted phenomenon, such as aggregation or coagulation, may occur during heating, consequently leading to an increase of particle size. As we can see in Figure 3.18, before heating, both catalysts reached almost the same value of selectivity. After 12 h at high temperature, the selectivity of the unmodified catalyst dropped twice as boron doped. Figure 3.19 summarizes the effect of adding boron to the catalyst.

The use of various materials and their impact on catalytic performance in dimethyl oxalate over copper-based catalysts will be discussed here. Zhu et al. [54] compared three different

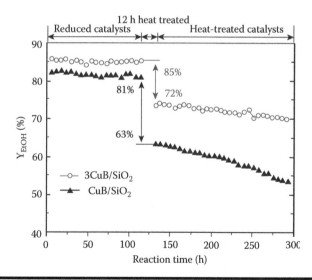

Figure 3.18   Catalytic properties of Cu/SiO_2 and $3CuB/SiO_2$ catalysts before and after treatment. (Reproduced from S. Zhao, H. Yue, Y. Zhao, B. Wang, Y. Geng, J. Lv, S. Wang, J. Gong, X. Ma, *J. Catal.* 297 (2013) 142–150. With permission.)

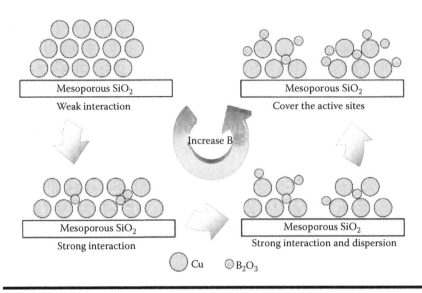

Figure 3.19 **Schematic model of the Cu species with increasing boron. (Reproduced from S. Zhao, H. Yue, Y. Zhao, B. Wang, Y. Geng, J. Lv, S. Wang, J. Gong, X. Ma, *J. Catal.* 297 (2013) 142–150. With permission.)**

Cu/oxide catalysts for this application. All ZrO_2, Al_2O_3, and SiO_2 were checked at the same conditions. Catalysts were prepared by a coprecipitation method and characterized by plenty of techniques. According to these results some morphologic differences can be observed. Overall, the scheme of these materials as the catalysts in the discussed process is illustrated in Figure 3.20.

From XRD patterns, a type of support stimulated growth of copper particles can be clearly seen. The smallest ones (21 nm) were detected in the Cu/ZrO_2 catalyst, which is obviously related with the highest copper dispersion. In contrast, the biggest particles were detected on a catalyst supported by silica. Additionally, the surface area for these two samples was quite different. Cu/SiO_2 indicated twice the surface area as Cu/ZrO_2. As before, the Cu^+/Cu^0 ratio had an enormous impact on selective C=O bond hydrogenation. The SiO_2 supported catalyst showed the lowest amount of Cu^0 species among samples,

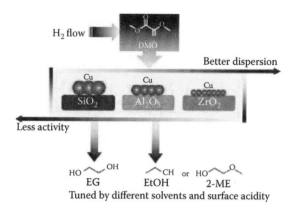

Figure 3.20 Catalytic performance on dimethyl oxalate hydrogenation over different supports. (Reproduced from Y. Zhu, Y. Zhu, G. Ding, S. Zhu, S. Zheng, Y. Li, *Appl. Catal. A* 468 (2013) 296–304. With permission.)

conversely to Cu/ZrO_2. It can be guessed that all these differences molded catalytic activity of tested materials (Figure 3.21).

As shown, Cu/SiO_2 indicates better selectivity to methyl glycol than others but solely during a relatively short time. Brief active period decreases the value of this catalyst in industry applications. On the other hand, great copper dispersion resulted in 100% conversion during the whole experiment. It is likely that selectivity was constant but at a lower level than in the initial phase of the experiment with silica. In reference to presented results, a proper combination of metal dispersion and Cu^+/Cu^0 seems to be the best way to achieve excellent conversion and selectivity in hydrogenation of dimethyl oxalate.

3.4 Concluding Remarks

Undoubtedly, there are many benefits of transition metal application in catalysis, such as cost and high selectivity, and copper is probably the best example of that. Because of a great ability to hydrogenate the C=O bond, even when C=C is

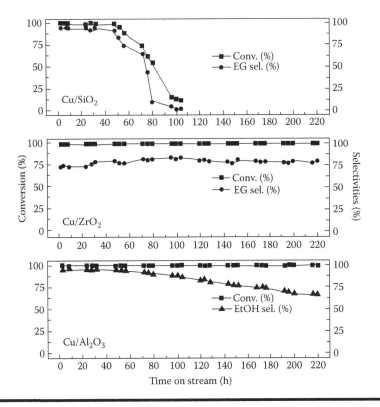

Figure 3.21 **Dimethyl oxalate (DMO) conversion and ethylene glycol (EG)/ethanol (EtOH) selectivity versus time on stream. (Reproduced from Y. Zhu, Y. Zhu, G. Ding, S. Zhu, S. Zheng, Y. Li, *Appl. Catal. A* 468 (2013) 296–304. With permission.)**

present in the same molecule, copper-based catalysts unveil a broad range of possibilities. As shown, plenty of fine chemicals with high industry meaning can be obtained by catalysis over this metal.

Nevertheless, manufacture in an industrial scale demands some specific properties from catalysts, and the current catalytic development, despite promising results at the laboratory scale, may not satisfy the industrial requirements. When catalysts are used industrially it is important that the shape and size of the particles provide a proper balance between activity in the process and selectivity to the desired product.

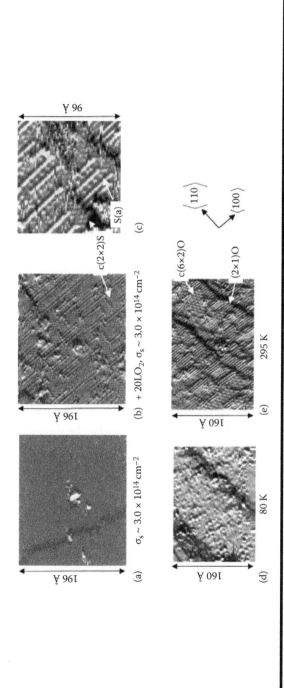

Figure 3.22 STM images of (a) disordered chemisorbed mobile sulfur adatoms (σ_s = 3 × 10^{14} cm^{-2}; θ = 0.28) at a Cu(110) surface at 22°C—note the resolved copper atoms in the (110) direction; (b) ordered structures of (2 × 1)—O and c(2 × 2)—S formed an exposure of (a) to 20 L oxygen at 22°C; (c) high-resolution image of chemisorbed sulfur adatoms separating (2 × 1)—O strings; (d) disordered oxygen states present at a Cu(110) surface at –193°C; (e) ordered (2 × 1)—O and (6 × 2)—O states present after warming (d) to 22°C (1 langmuir [L] ≡ 10^{-6} Torr s). (Reproduced from M. V. Twigg, M. S. Spencer, *Appl. Catal. A* 212 (2001) 161–174. With permission.)

Moreover, catalysts should operate continuously, which makes them more susceptible to a decrease in performance as a consequence of, for example, poisoning. If material is vulnerable to deactivation, the materials need to allow for regeneration procedures. In industry, catalyst regeneration is carried out in a parallel reactors so the process is not interrupted. To minimize or totally eliminate these operations, the catalyst used should exhibit high stability. Thus, all these aspects have to be considered during the catalyst design process, in particular what kind of regeneration procedures can be applied—especially when copper is supported on materials that cannot be submitted to heat treatments, which is the common regeneration procedure at an industrial scale. Studies on catalysts deactivation are not widely available and will play an important role in design and developing the catalysts of the future (Figure 3.22) [60].

As mentioned before, the interaction between metal and support plays a significant role in its possible applications. It was also verified that different types of support might stimulate catalytic activity by particle size control to varying degrees. Additionally, a stronger metal-support effect has meaningful impact on catalyst stability. According to these factors, proper metal and novel carrier combination seems to be key in catalyst development.

References

1. T. Ohkuma, R. Noyori, *Comprehensive asymmetric catalysis,* vol. 1, E. N. Jacobsen, A. Pfaltz, H. Yamamo (eds.) Springer, Berlin-Heidelberg (1999) 199–246.
2. T. Ohkuma, R. Noyori, *Transition metals in organic synthesis,* vol. 2, M. Beller, K. Bolm (eds.) Wiley-VCH, Weinheim (1998) 29–41.
3. Y. Chi, W. Tang, X. Zhang, *Modern rhodium catalyzed organic reactions,* P. A. Evans (ed.) Wiley-VCH, Weinheim (2005) 1–31.

4. R. M. Bullock, An iron catalyst for ketone hydrogenations under mild conditions. *Angew. Chem. Int. Ed.* 46 (2007) 7360–7363.

5. S. Enthaler, K. Junge, M. Beller, Sustainable metal catalysis with iron: From rust to a rising star? *Angew. Chem. Int. Ed.* 47 (2008) 3317–3321.

6. S. Gaillard, J.-L. Renaud, Iron-catalyzed hydrogenation, hydride transfer, and hydrosilylation: An alternative to precious-metal complexes? *ChemSusChem* 1 (2008) 505–509.

7. B. Plietker, *Iron catalysis in organic chemistry*, 1st ed., Wiley-VCH, Weinheim (2008).

8. R. H. Morris, Asymmetric hydrogenation, transfer hydrogenation and hydrosilylation of ketones catalyzed by iron complexes. *Chem. Soc. Rev.* 38 (2009) 2282–2291.

9. K. Junge, B. Wendt, D. Addis, S. Zhou, S. Das, S. Fleischer, M. Beller, Copper-catalyzed enantioselective hydrogenation of ketones. *Chem. Eur. J.* 17 (2011) 101–105.

10. J. Gajewy, M. Kwit, J. Gawrońskia, Convenient enantioselective hydrosilylation of ketones catalyzed by zinc-macrocyclic oligoamine complexes. *Adv. Synth. Catal.* 351 (2009) 1055–1063.

11. S. Das, D. Addis, S. Zhou, K. Junge, M. Beller, Zinc-catalyzed reduction of amides: Unprecedented selectivity and functional group tolerance. *J. Am. Chem. Soc.* 132 (2010) 1770–1177.

12. N. A. Marinos, S. Enthaler, M. Driess, High efficiency in catalytic hydrosilylation of ketones with zinc-based precatalysts featuring hard and soft tridentate O,S,O-ligands. *ChemCatChem* 2 (2010) 846–852.

13. P. Sabatier, C. Senderens, *Ann. Chim. Phys.* 4 (1905) 368.

14. R. N. Pease, R. B. Purdum, The hydrogenation of benzene in the presence of metallic copper. *J. Am. Chem. Soc.* 47 (1925) 1435–1442.

15. V. N. Ipatieff, B. B. Corson, I. D. Kurbatov, Copper as catalyst for the hydrogenation of benzene. *J. Phys. Chem.* 43 (1939) 589–604.

16. M. R. Decan, S. Impellizzeri, M. L. Marin, J. C. Scaiano, Copper nanoparticle heterogeneous catalytic 'click' cycloaddition confirmed by single-molecule spectroscopy. *Nat. Commun.* 5 (2014) 4612.

17. H. Bönnemann, R. M. Richards, Nanoscopic metal particles — synthetic methods and potential applications. *Eur. J. Inorg. Chem.* 2001 (2001) 2455–2480.

18. C. Huang, H. Zhang, Y. Zhao, S. Chen, Z. Liu, Diatomite-supported Pd–M (M = Cu, Co, Ni) bimetal nanocatalysts for selective hydrogenation of long-chain aliphatic esters. *J. Coll. Interface Sci.* 386 (2012) 60–65.

19. T. M. D. Dang, T. T. T. Le, E. Fribourg-Blanc, M. C. Dang, Synthesis and optical properties of copper nanoparticles prepared by a chemical reduction method. *Adv. Nat. Sci. Nanosci. Nanotechnol.* 2 (2011) 015009.

20. M. B. Boucher, B. Zugic, G. Cladaras, J. Kammert, M. Marcinkowski, T. J. Lawton, E. C. H. Sykes, M. Flytzani-Stephanopoulos, Single atom alloy surface analogs in $Pd_{0.18}Cu_{15}$ nanoparticles for selective hydrogenation reactions. *Phys. Chem. Chem. Phys.* 15 (2013) 12187–12196.

21. C. Wu, B. P. Mosher, T. Zeng, One-step green route to narrowly dispersed copper nanocrystals. *J. Nanoparticle Res.* 8 (2006) 965–969.

22. S. Mandal, S. De, Catalytic and fluorescence studies with copper nanoparticles synthesized in polysorbates of varying hydrophobicity. *Coll. Surf. A Physicochem. Eng. Asp.* 467 (2015) 233–250.

23. N. A. Dhas, C. P. Raj, A. Gedanken, *Properties of Metallic*. 4756 (1998) 1446–1452.

24. V. Gutiérrez, F. Nador, G. Radivoy, M. A. Volpe, Highly selective copper nanoparticles for the hydrogenation of α,β-unsaturated aldehydes in liquid phase. *Appl. Catal. A* 464–465 (2013) 109–115.

25. M. Blosi, S. Albonetti, M. Dondi, C. Martelli, G. Baldi, Microwave-assisted polyol synthesis of Cu nanoparticles. *J. Nanoparticle Res.* 13 (2010) 127–138.

26. Y. Zhao, J. J. Zhu, J.-M. Hong, N. Bian, H.-Y. Chen, Microwave-induced polyol-process synthesis of copper and copper oxide nanocrystals with controllable morphology. *Eur. J. Inorg. Chem.* 2004 (2004) 4072–4080.

27. H. Kawasaki, Y. Kosaka, Y. Myoujin, T. Narushima, T. Yonezawa, R. Arakawa, Microwave-assisted polyol synthesis of copper nanocrystals without using additional protective agents. *Chem. Commun.* 47 (2011) 7740–7742.

28. Y. H. Kim, D. K. Lee, B. G. Jo, J. H. Jeong, Y. S. Kang, Synthesis of oleate capped Cu nanoparticles by thermal decomposition. *Coll. Surf. A Physicochem. Eng. Asp.* 284–285 (2006) 364–368.

29. M. Salavati-Niasari, N. Mir, F. Davar, A novel precursor for synthesis of metallic copper nanocrystals by thermal decomposition approach. *Appl. Surf. Sci.* 256 (2010) 4003–4008.
30. M. Salavati-Niasari, F. Davar, N. Mir, Synthesis and characterization of metallic copper nanoparticles via thermal decomposition. *Polyhedron* 27 (2008) 3514–3518.
31. S. Wang, X. Li, Q. Yin, L. Zhu, Z. Luo, Highly active and selective Cu/SiO_2 catalysts prepared by the urea hydrolysis method in dimethyl oxalate hydrogenation. *Catal. Commun.* 12 (2011) 1246–1250.
32. R. Betancourt-Galindo, P. Y. Reyes-Rodriguez, B. A. Puente-Urbina, C. A. Avila-Orta, O. S. Rodrígues-Fernández, G. Cadenas-Pliego, R. H. Lira-Saldivar, L. A. García-Cerda, Synthesis of copper nanoparticles by thermal decomposition and their antimicrobial properties. *J. Nanomater.* 2014 (2014) 1–5.
33. H. Hashemipour, M. E. Zadeh, R. Pourakbari, P. Rahimi, Investigation on synthesis and size control of copper nanoparticle via electrochemical and chemical reduction method. *Int. J. Phys. Sci.* 6 (2011) 4331–4336.
34. P. Gallezot, D. Richard, Selective hydrogenation of α,β-unsaturated aldehydes. *Catal. Rev. Sci. Eng.* 40 (1998) 81–126.
35. R. Zheng, M. Porosoff, J. Weiner, S. Lu, Y. Zhu, J. Chen, Controlling hydrogenation of C=O and C=C bonds in cinnamaldehyde using silica supported Co-Pt and Cu-Pt bimetallic catalysts. *Appl. Catal. A* 419–420 (2012) 126–132.
36. A. J. Marchi, D. A. Gordo, A. F. Trasarti, C. R. Apesteguia, Liquid phase hydrogenation of cinnamaldehyde on Cu-based catalysts. *Appl. Catal. A* 249 (2003) 53–67.
37. Z. Liu, Y. Yang, J. Mi, X. Tan, Y. Song, Synthesis of copper-containing ordered mesoporous carbons for selective hydrogenation of cinnamaldehyde. *Catal. Commun.* 21 (2012) 58–62.
38. A. Chambers, S. D. Jackson, D. Stirling, G. Webb, Selective hydrogenation of cinnamaldehyde over supported copper catalysts. *J. Catal.* 168 (1997) 301–314.
39. S. Valange, J. Barrault, A. Derouault, Z. Gabelica, Preparation of nickel-mesoporous materials and their application to the hydrodechlorination of chlorinated organic compounds. *Micro. Meso. Mater.* 211 (2001) 44–45.

40. W. Jong, G. Marcotullio, Overview of biorefineries based on co-production of furfural, existing concepts and novel developments. *Int. J. Chem. React. Eng.* 8 (2010) A69.

41. I. Horváth, H. Mehdi, V. Fábos, L. Boda, L. Mika, γ-valerolactone—A sustainable liquid for energy and carbon-based chemicals. *Green Chem.* 10 (2008) 238–242.

42. Y. Yao, Z. Wang, S. Zhao, D. Wang, Z. Wu, M. Zhang, A stable and effective Ru/polyethersulfone catalyst for levulinic acid hydrogenation to γ-valerolactone in aqueous solution. *Catal. Today* 234 (2014) 245–250.

43. M. Sudhakar, M. Lakshmi Kantam, V. Swarna Jaya, R. Kishore, K. V. Ramanujachary, A. Venugopal, Hydroxyapatite as a novel support for Ru in the hydrogenation of levulinic acid to γ-valerolactone. *Catal. Commun.* 50 (2014) 101–104.

44. W. Luo, U. Deka, A. Beale, E. van Eck, P. Bruijnincx, B. Weckhuysen, Ruthenium-catalyzed hydrogenation of levulinic acid: Influence of the support and solvent on catalyst selectivity and stability. *J. Catal.* 301 (2013) 175–186.

45. P. Upare, M. Jeong, Y. Hwang, D. Kim, Y. Kim, D. Hwang, U. Lee, J. Chang, Development of hierarchical EU-1 zeolite by sequential alkaline and acid treatments for selective dimethyl ether to propylene (DTP). *Appl. Catal. A* 491 (2015) 127–135.

46. B. Putrakumar, N. Nagaraju, V. Kumar, K. Chary, Hydrogenation of levulinic acid to γ-valerolactone over copper catalysts supported on γ-Al_2O_3. *Catal. Today* 250 (2015) 209–217.

47. I. Obregón, E. Corro, U. Izquierdo, J. Requies, P. Arias, Levulinic acid hydrogenolysis on Al_2O_3-based Ni-Cu bimetallic catalysts. *Chin. J. Catal.* 35 (2014) 656–662.

48. B. Nagaraja, V. Kumar, V. Shasikala, A. Padmasri, B. Sreedhar, B. Raju, K. Rao, A highly efficient Cu/MgO catalyst for vapour phase hydrogenation of furfural to furfuryl alcohol. *Catal. Commun.* 4 (2003) 287–293.

49. B. Nagaraja, A. Padmasri, B. Raju, K. Rao, Vapor phase selective hydrogenation of furfural to furfuryl alcohol over Cu–MgO coprecipitated catalysts. *J. Mol. Catal. A* 265 (2007) 90–97.

50. D. Hernández, J. Caballero, J. González, R. Tost, J. Robles, M. Cruz, A. López, R. Huesca, P. Torres, Furfuryl alcohol from furfural hydrogenation over copper supported on SBA-15 silica catalysts. *J. Mol. Catal. A* 383–384 (2014) 106–113.

51. M. Villaverde, N. Bertero, T. Garetto, A. Marchi, Selective liquid-phase hydrogenation of furfural to furfuryl alcohol over Cu-based catalysts. *Catal. Today* 213 (2013) 87–92.

52. M. Villaverde, T. Garetto, A. Marchi, Liquid-phase transfer hydrogenation of furfural to furfuryl alcohol on Cu–Mg–Al catalysts. *Catal. Commun.* 58 (2015) 6–10.

53. H. Yue, Y. Zhao, X. Ma, J. Gong, Ethylene glycol: Properties, synthesis, and applications. *Chem. Soc. Rev.* 41 (2012) 4218–4244.

54. Y. Zhu, Y. Zhu, G. Ding, S. Zhu, S. Zheng, Y. Li, Highly selective synthesis of ethylene glycol and ethanol via hydrogenation of dimethyl oxalate on Cu catalysts: Influence of support. *Appl. Catal. A* 468 (2013) 296–304.

55. A. Yin, X. Guo, W. Dai, H. Li, K. Fan, Highly active and selective copper-containing HMS catalyst in the hydrogenation of dimethyl oxalate to ethylene glycol. *Appl. Catal. A* 349 (2008) 91–99.

56. C. Wen, Y. Cui, X. Chen, B. Zong, W. Dai, Reaction temperature controlled selective hydrogenation of dimethyl oxalate to methyl glycolate and ethylene glycol over copper-hydroxyapatite catalysts. *Appl. Catal. B* 162 (2015) 483–493.

57. L. Chen, P. Guo, M. Qiao, S. Yan, H. Li, W. Shen, H. Xu, K. Han, Cu/SiO_2 catalysts prepared by the ammonia-evaporation method: Texture, structure, and catalytic performance in hydrogenation of dimethyl oxalate to ethylene glycol. *J. Catal.* 257 (2008) 172–180.

58. A. Yin, X. Guo, W. Dai, K. Fan, Effect of initial precipitation temperature on the structural evolution and catalytic behavior of Cu/SiO_2 catalyst in the hydrogenation of dimethyloxalate. *Catal. Commun.* 12 (2011) 412–416.

59. S. Zhao, H. Yue, Y. Zhao, B. Wang, Y. Geng, J. Lv, S. Wang, J. Gong, X. Ma, Chemoselective synthesis of ethanol via hydrogenation of dimethyl oxalate on Cu/SiO_2: Enhanced stability with boron dopant. *J. Catal.* 297 (2013) 142–150.

60. M. V. Twigg, M. S. Spencer, Deactivation of supported copper metal catalysts for hydrogenation reactions. *Appl. Catal. A* 212 (2001) 161–174.

Chapter 4

Hydrogenation by Iron Catalysts

Tomiła Łojewska and Jacinto Sá

Contents

This chapter deals with hydrogenation mediated by iron metal present as supported and unsupported nanoparticles. Inspired by nature, design scientists have developed homogeneous and heterogeneous systems to perform selective hydrogenation. In the chapter, we highlight synthetic procedures for the preparation of metallic iron nanoparticles and their applications to fine and pharmaceutical chemical production.

4.1 Introduction—Iron, the Natural Choice

As learned from evolution, mimicking biological systems is proven to be a good strategy, which gives an array of ingenious solutions in many areas of science. In this sense, iron seems to be a natural choice for hydrogenation catalysis since there are numerous examples of natural systems that have iron incorporated in their catalytic processes, some of them capable of hydrogen activation. In particular, certain microbial species are capable of breathing with hydrogen thanks to enzymes—hydrogenases—the majority of which have iron cations in their active sites. Their complex structure has been investigated since the 1930s [1] and certain motifs became an inspiration for homogeneous catalysts. It is also worth mentioning that hydrogenases are valuable models for catalytic hydrogen generation. Hydrogen fueled cars and buses have recently been released into the consumer market, which means that clean and benign hydrogen production technologies to replace fossil fuels are in high demand today. Therefore, the number of publications dealing with hydrogen as an alternative energy carrier or a reductant for hydrogenation/ reduction reactions has increased rapidly in the last few years.

Hydrogenases are iron-containing metalloenzymes, found in some microbial organisms that use hydrogen as an energy source. The reaction is one of the steps of different metabolic pathways and it is accompanied by some other redox reactions of inorganic entities as is schematically shown in Figure 4.1. Hydrogenases are of great interest due to the underlying successful strategy for obtaining impressive performances of hydrogen production and activation. Indeed, the hydrogenases have turnover frequency (TOF) of the order of 10,000 s^{-1} [2], unmatched by any other catalytic system.

There are a variety of hydrogenases widespread in many different species of bacteria, archaea, and prokaryotes. In terms of chemical composition, they can be divided into three classes: [FeFe], [FeNi], and [Fe-only] hydrogenases. Their active sites as well as the protein matrix are very complex, especially regarding the seemingly simple reaction (hydrogen production/breakup) it is responsible for. Figure 4.1 shows a possible mechanism of a [Fe-Fe] hydrogenase that highlights system complexity. This is out of the scope of this chapter, so readers are advised to read the Lubitz et al. [3] review for further information.

The very fact that nature has developed a way to use iron is perhaps determined to some extent by the sheer abundance of this element, which accounts for 6% of the Earth's crust [4]. Likewise, the low price of iron is now also a driving force to effectively exploit iron as a catalyst. Yet, using it

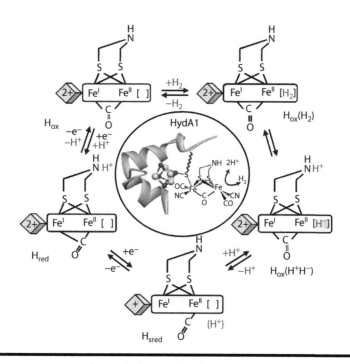

Figure 4.1 Proposed catalytic mechanism of [FeFe] hydrogenase including the H$_{sred}$ state. (H⁺) in H$_{sred}$ indicates that the proton is believed to be associated with the H-cluster and possibly bound to a nearby amino acid residue. (Reproduced from W. Lubitz, H. Ogata, O. Rüdiger, E. Reijerse, *Chem. Rev.* 114 (2014) 4081. With permission.)

still remains a challenge. It is not without reason that, up to the present, iron has been neglected in favor of other metals, such as Pt, Pd, or Rh, which have earned their names as traditional hydrogenation catalysts. To take advantage of iron's catalytic activity, we have to learn how to overcome the main obstacle, which is the extreme reactivity and pyrophoricity of iron nanoparticles.

Coming up with a viable solution to circumvent the shortcomings of iron as a hydrogenation catalyst would be very rewarding. Apart from the economic and availability factors, there are some additional reasons, namely:

a. Iron is a greener metal compared to traditionally used heavy metals.
b. Iron is the least toxic metal.
c. Iron exhibits unique magnetic properties that simplify separation and potential enhanced activities [5,6].

To date there have been some attempts to tackle the vexing problem of iron utilization as mentioned earlier; they have often been successful, although this area of research is rather in its developmental stage. It thus seems quite plausible that in the next decade maybe we will witness the forecasted "new iron age" [7].

Using iron in catalysis, especially as a precious metal replacement, is not a new idea. In 1909, Mittasch at BASF was charged with finding more economical catalysts as replacements for existing osmium and uranium compounds used in industrial ammonia synthesis. An activated iron surface was discovered that became the foundation for modern Haber–Bosch technology [8]. Other early applications of iron catalysis include Gif chemistry, the Reppe cyclization, biomimetic chemistry, Lewis acid promoted reactions, and the foundations of modern cross-coupling methods [9]. Global sustainability efforts, coupled with the observation of unique and interesting reactivity, have resulted in a recent renaissance in iron catalysis. Transformations including cycloadditions, cross couplings, oxidations, and olefin polymerizations are currently the subject of intense investigations from many different academic and industrial research groups [7].

This chapter will provide a short historical perspective on iron catalysis with focus on the prospects of using the experience of homogeneous and bio-iron catalysts in heterogeneous catalysis for fine chemical industry. Moreover, we will present current approaches toward mitigating the problems with practical application of zero-valent iron.

4.2 Nanoparticle Synthesis

Until recently, catalytic applications of nano-Fe^0 have been hindered due to inconvenient and/or ineffective synthetic routes giving difficult to handle, phyrophoric nanocrystals, of poor quality, and, most importantly, of broad size distribution [4]. Control of crystal morphology is of great relevance in catalysis not only because of the active surface area but also due to the structure sensitive reaction nature of most chemical transformations.

The high degree of polydispersity stems from the tendency of nano-Fe^0 to agglomerate, which is also true for any other magnetic nanoparticles, like Co or Fe oxides. This is due to van der Waals forces and magnetic attraction [10], which make such nanoparticles vulnerable to aggregation. Currently, the synthesis of stable iron nanoparticles has advanced considerably, especially because of very promising prospects of using zero-valent iron in environmental remediation technologies. Iron was recently proved to be an efficient agent for transformation and detoxification of many contaminants. The idea is to take advantage of iron's extreme reactivity (which is at the same time nontoxic and inexpensive) for in situ soil, aquifer, and wastewater remediation by direct injection or in permeable reactive barriers (PRBs) [11].

There is a wide array of nano-Fe^0 preparation methods that follow the same classification as in any other metal nanoparticles (NP) synthesis—namely, bottom-up and top-down approaches. The bottom-up approach is usually favored over the top-down techniques, like mechanical milling, which give relatively big nanoparticles (100 nm), with broad size distribution and are difficult to scale up to support industrial applications [12]. Furthermore, top-down methods are usually not very well suited to prepare shaped particles and very small sizes [13]. Table 4.1 shows the general methods used for the preparation of nano-Fe^0.

As stated before, any zero-valent iron synthesis procedure, especially for catalysis applications, must be tailored according to the high chemical activity, as well as the tendency to oxidize on air. Thus, different strategies have been incorporated to classical synthetic methodologies in order to minimize the rapid passivation or self-ignition of the active nanoparticles. Common requirements include inert reaction and handling conditions (e.g., glove boxes or Schlenk systems), in particular dry solvents and reagents or reductive atmosphere. An example of such a synthesis is the work by Zhang et al. [14], who conducted thermal decomposition of iron acethylacetonate under high-pressure hydrogen atmosphere with trioctylphosphine and octylamine used as stabilizing ligands. The x-ray and electron diffraction patterns confirmed that they produce pure α-iron. Electron microscopy revealed uniform

Table 4.1 General Methods for the Preparation of Nano-Fe^0

Bottom-Up	Top-Down
Thermal decomposition	Milling
Chemical reduction of Fe salts/complexes	Vapor phase
Green methods	
Microemulsion	
Sonification of Fe compounds	

nanoparticles with an average diameter of 11.5 nm and a superlattice of body-centered cubic (bcc)-Fe with a high degree of crystallinity.

Thermal decomposition is a popular method for fabrication of iron nanoparticles; however, standard procedure typically involves iron pentacarbonyl as the metal source, since it decomposes easily at low temperatures and does not give by-products [4]. This method is usually applied in the presence of high-boiling-point organic solvents and stabilizing surfactants/ligands [15]. This preparation route yields mostly amorphous iron nanoparticles [16]—although Lacroix et al. [17] recently described a synthetic methodology giving uniform nanocrystals with bcc lattice structure and high magnetization (Figure 4.2). In their work a surfactant in a postsynthetic ligand exchange procedure was used. However, in some examples the formed iron nanoparticles were found to have a relatively thin Fe_3O_4 passivation shell that made them stable in water solutions but can potentially hinder their application to hydrogenation catalysis.

Iron pentacarbonyl can also be readily decomposed into iron nanoparticles by means of ultrasonic radiation in the presence of stabilizing agents. Suslick et al. [18,19] introduced such a sonochemical approach for the synthesis of iron nanoclusters, which was subsequently applied for the production of plethora metal and other nanoparticles. It is worth emphasizing that in the absence of stabilizing ligands, yields in amorphous iron (powder) are prone to aggregation [20], whereas by cavitation of pentacarbonyl with silica, supported iron can be obtained [21–23].

Since iron pentacarbonyl is highly toxic, chemical reduction methods that incorporate iron salts or oxides are preferred because they are environmentally friendlier. Successful synthesis of iron nanoparticles utilizing tea leaf [24], alfalfa biomass [25], sorghum bran extracts [26], and fungi [27] has been reported recently. The synthesis of iron nanoparticles using extracts from biological sources is significantly easier as

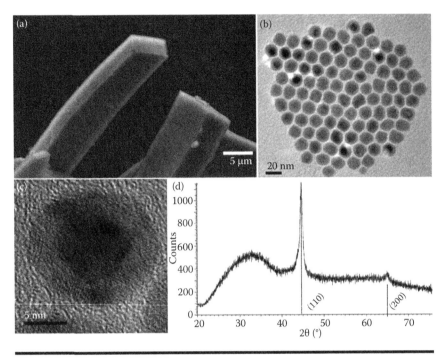

Figure 4.2 **Morphological and structural characterization of bcc-Fe/ Fe$_3$O$_4$ nanoparticles. (a) SEM image of the plate-like Fe nanoparticles assembly obtained directly from the synthesis solution. (b) TEM image of the 15 nm NPs obtained from the redispersion of the plate assembly in hexanes. (c) HRTEM image of a single nanoparticle revealing the metallic bcc-Fe core and Fe$_3$O$_4$ shell with Fe (110) and Fe$_3$O$_4$ (222) planes indicated. (d) XRD pattern of the bcc-Fe nanoparticles. (Reproduced from L.-M. Lacroix, N. F. Huls, D. Ho, X. Sun, K. Cheng, S. Sun, *Nano Lett.* 11 (2011) 1641. With permission.)**

compared to conventional chemical methods. For example, plant iron salt solutions can simply be added to the plant extract solution with an optimum volume ratio at ambient temperatures and pressures. The utilization of natural or organic acids is also an alternative way for iron nanoparticle preparation. For example, Meeks et al. used ascorbic acid for the synthesis of iron-based nanoparticles [28].

Another important route to prepare iron nanoparticles is the use of microorganisms such as bacteria, fungi, and yeast [29,30]. Recently, Bharde et al. [31] reported the preparation of nanomagnetite within 24 h with *Actinobacter* spp. starting from an aqueous potassium ferricyanide/ferrocyanide mixture under aerobic conditions (Figure 4.3). These microbes yield nanoparticles with fine morphology through either intracellular or extracellular mechanisms. Based on reducing or precipitating soluble toxic metal ions, microorganisms generate insoluble nontoxic metal nanoclusters [32]. The advantages of green reducing agents are that they are environmentally benign and the products obtained are both stable and controllable.

Another solution to attain monodisperse nanoparticles is seeding with another metal such as palladium ions. Huang and Ehrman [33] demonstrated that in the presence of

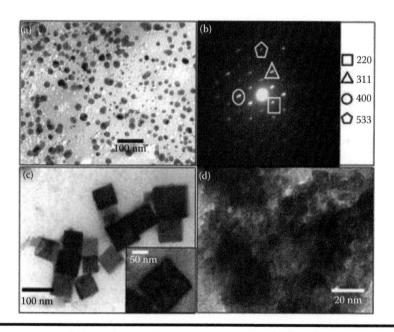

Figure 4.3 **(a) TEM image of magnetite nanoparticles after 24 h of reaction with *Actinobacter* spp. (b) SAED pattern of the particles in (a). (c and inset) TEM images of magnetite nanoparticles after 48 h of reaction with *Actinobacter* spp. (d) TEM image of the 48 h reacted magnetite nanoparticles after calcination at 350°C for 3 h. (Reproduced from A. Bharde, A. Wani, Y. Shouche, P. A. Joy, B. L. V. Prasad, M. Sastry, *J. Am. Chem. Soc.* 127 (2005) 9326. With permission.)**

palladium ions and controlled pH, they obtained well mono-dispersed iron nanoparticles. The effect of pH in the synthesis is schematically illustrated in Figure 4.4. Combined with seeding, the use of templates made of surfactants can help control the size, shape, and polydispersity of nanosized metal particles even further [34].

In an attempt to automate the procedure of iron nanoparticle production, Bedford et al. [35] developed a carousel reactor system. The iron nanoparticles were formed in situ stabilized by 1,6-bis(diphenylphosphino)hexane or polyethylene glycol (PEG). The resultant nanomaterials were found to be excellent catalysts for the cross coupling of aryl Grignard reagents with primary and secondary alkyl halides bearing β-hydrogens and they also proved effective in a tandem cyclization/cross-coupling reaction.

Mechanical grinding of bulk metal is admittedly a feasible nano-iron fabrication method for large-scale industrial applications, and it has already been made commercially available in large quantities [36,37]. However, it is not suitable for heterogeneous catalysis because of the aforementioned polydispersity of NPs obtained in this way.

All the techniques described provide colloidal zero-valent iron nanoparticles. Thus, it is understandable that in order to utilize them in heterogeneous catalysis, they should be embedded on some kind of a support. This has proved to work in the case of other metal catalysts. For example, palladium

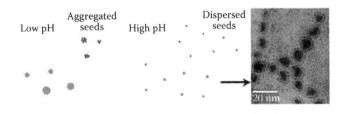

Figure 4.4 **Effect of pH in Fe nanoparticles formation using palladium ions as seeds. (Reproduced from K. Huang, S. H. Ehrman,** *Langmuir* **34 (2007) 1419. With permission.)**

nanoparticles grafted to a polystyrene–poly(ethylene glycol) (PS-PEG) block copolymer gave high catalytic activity in the hydrogenation of olefins and the hydrodechlorination of chloroarenes under aqueous conditions [38]. Because immobilization of iron nanoparticles was also required for water/ soil remediation techniques developed during the last decade, several solutions can be implemented and then optimized to meet the specific demands of a catalytic application. This is summarized in Table 4.2.

Finally, an approach for zero-valent iron hydrogenation catalyst fabrication is to reduce the presynthesized Fe-oxide catalyst, as proposed for the Fischer–Tropsch process [12,39–41]. The approach stabilizes and protects the morphology of the nanoparticles, but it requires harsh reduction conditions, which can damage the initial morphology.

4.3 Hydrogenation Catalysis

The use of iron as an alternative to noble metals is not a new idea. The most prominent uses of iron in heterogeneous catalysis are the Haber–Bosch (ammonia synthesis from nitrogen and hydrogen) and Fischer–Tropsch (conversion of CO and hydrogen into hydrocarbons) processes. Even though their introduction into large-scale industrial use (e.g., nowadays the Haber process is said to account for 1% of world energy consumption) dates back to the beginning of the twentieth century, the structure–reactivity relationships in the processes mentioned are still the subject of hot debates. From the perspective of the chemical industry there is still a great demand for improving the existing catalysts and designing more sustainable systems, especially with the aid of current nanotechnological achievements. Some most recent examples are cited in References [40,42,43]; however, it should be taken into account that the most crucial achievements in this field are kept secret among the big companies. In fact, new developments

Table 4.2 Supports for Fe Nanoparticles: Their Roles and Applications

Support Type	Support Examples	Role of the Support	Intended Application of Fe Nanoparticles/Composite	Ref.
Polymers	Polyacrylic acid (PAA)	Prevents aggregation of nanoparticles	Small Fe nanoparticles (6 nm)—for general purpose	[33]
	Poly(2,6-dimethyl-1,4-phenyleneoxide)	Stabilization	Very small Fe nanoparticles with preserved magnetic properties	[44]
	Polyaniline nanofibers	Stabilization, immobilization	Adsorbent of As from aqueous solutions	[45]
	Polystyrene/(polyethylene glycol)	Immobilization, stabilization	Hydrogenation in water flow	[46]
Porous solids	Zeolite (natural)	Support, absorbent	Degradation of dyes catalyzed by Fe nanoparticles	[47]
	Mercaptopropyl-modified silica	Support, absorbent	CO_2 absorption	[48]

(Continued)

Table 4.2 (Continued)   Supports for Fe Nanoparticles: Their Roles and Applications

Support Type	Support Examples	Role of the Support	Intended Application of Fe Nanoparticles/Composite	Ref.
Functionalized clays	Mg-amino clay	Prevents aggregation and enables tuning the size of synthesized nanoparticles, increasing reactivity of Fe	Remediation of subsurface contaminants (e.g., nitrates)	[10]
Graphene	Graphene composite	Supporting matrix	Uranium removal from groundwater	[49]
	Fe_3O_4/graphene nanocomposite		Chromium(VI) removal	[50]
Cellulose	Cellulose fibers (filter paper)	Immobilization, increasing activity toward phenols to catechols transformation	Magnetic filter paper for water treatment (e.g., removal of Cr(VI) or phenols)	[51]
	Carboxymethyl cellulose (CMC)	Stabilization and transportation of nanoparticles in porous media (soils)	Water decontamination	[52]

addressing industrial uses of iron are not in the scope of this book, as we will focus on the slowly emerging examples of iron nanocatalysts for potential hydrogenation reactions applied in the synthesis of fine chemical precursors. Also, we purposely exclude an array of bimetallic catalysts with zero-valent iron where it has been revealed that iron is completely inactive [53,54] or much less active in comparison with the other metal used [55].

Heterogeneous nanocatalysis concepts for effective hydrogenation have been inspired from homogenous catalysis with iron salts or complexes. As in the case of any other transition metals discussed in this book, the highly desired, primordial goal is to find the correlation between the structure of nanomaterials influenced by the external environment (surrounding scaffold, some additives being used, etc.) and the catalytic activity. Using nanoparticles as active agents is in fact a frontier between the two fields: homogeneous and heterogeneous catalysis. Sometimes, to realize the difference between these two realms is not an obvious issue to consider—the example of which is the work by Sonneberg et al. [56]. The work was devoted to determining the nature of the catalyst activity centers by proving that the transfer hydrogenation reaction of ketones was activated by zero-valent iron nanoparticles and not by soluble iron species. The authors admitted that they could not exclude the influence of homogeneous, iron-containing agents as being present in the reaction network.

Selectivity is a crucial factor for the production of fine chemicals, which is traditionally associated with homogeneous catalysis. Careful design of iron carbonyl, phosphine, and other complexes allowed for asymmetric hydrogenation and enantioselective hydrosilylation of many functional compounds [57–59]. But on the other hand, the two key attributes of heterogeneous nanocatalysts—the big surface-to-volume ratio, which accounts for the superior performance of nanomaterials and easy separation—favor their application in the future. Additionally, recent papers report that some degree of

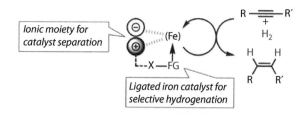

Figure 4.5 Modular ion pair/ligand/iron catalyst for hydrogenation. (Reproduced from T. N. Gieshoff, A. Welther, M. T. Kessler, M. H. G. Prechtl, A. J. von Wangelin, *Chem. Commun.* 50 (2014) 2261. With permission.)

chemo- or stereoselectivity of iron nanoparticles can be possible. One of the resolved approaches to this is a specially designed ligand-modified catalyst [60], which combines the homo- and heterogeneous strategies (Figure 4.5). Again, the distinction between the two realms was not obvious.

The separation aspect makes the application of unsupported colloidal nanoparticles limited because the catalyst is difficult to recycle, and the morphology and size distribution of the as-prepared nanoparticles change as reaction proceeds. These call for better understanding and more effective control of morphology of colloidal supported catalysts. Still, the research area is in its infancy and the catalytic potential of zero-valent iron nanomaterials has not yet been fully exploited.

4.3.1 Hydrogenation of Alkenes

Phua et al. [61] used soluble iron nanoparticles to hydrogenate C=C unsaturated bonds, which they prepared via a variant of a chemical reduction method, proposed by Bedford et al. [35], in which the Grignard reagent, EtMgCl, is used to reduce $FeCl_3$, yielding nanoparticles with ~2.5 nm in diameter, supposedly stabilized by $MgCl_2$ ions (Figure 4.6). The as-prepared nanomaterial was tested for hydrogenation of substituted and non-substituted, aliphatic, and cyclic olefins. The results showed

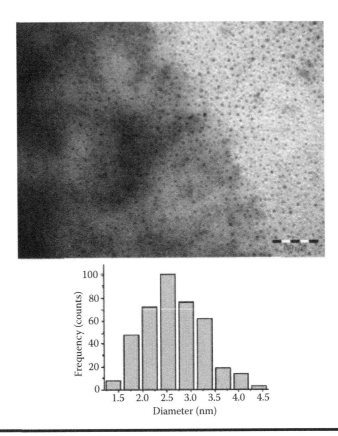

Figure 4.6   Iron nanoparticles obtained via Grignard reduction. TEM micrograph and histogram showing nanoparticles size distribution. (Reproduced from P.-H. Phua, L. Lefort, J. A. F. Boogers, M. Tristany, J. G. de Vries, *Chem. Commun.* 25 (2009) 3747. With permission.)

that full conversion was achieved in mild conditions (even at room temperature) for terminal substrates, whereas the catalyst exhibits no activity toward either hydrogenation of tri- and tetra-substituted or internal alkenes. In this way, such catalysts can offer selectivity when more than one type of unsaturated bond is present in a molecule. Cyclic (nonsubstituted) olefins required a higher temperature of 100°C. However, best hydrogenation performance was observed for norbornene, where full conversion was attained at room temperature and only 1 bar pressure of molecular hydrogen.

The recyclability studies showed that the catalyst can be used in up to five reaction cycles without significant loss in activity, which indicates that this nanoparticle stabilization is effective. However, as for the stability of the colloidal iron catalyst toward the exposure to water and air, the reverse is also true. Even as little as 1% (v/v) of water inhibited the reaction almost completely. Another considerable drawback is that THF, which was the solvent used here, is not a desirable and environmentally friendly medium for larger scale applications.

The same group carried out thorough investigation on the hydrogenation reaction parameters, including the influence of the Bedford-inspired catalyst synthesis conditions [35]. Several factors were considered: Namely, the reducing agent having been chosen, iron precursor, solvent and stabilizers, and their impact were tested for norbornene hydrogenation. The results have shown that the type of organometallic reducer—namely, trialkyl aluminums, Grignard reagents, tertra-alkyllead, and tetra-alkyltin—were indeed critical. While, for example, Me_3Al gave no product at all, Et_3Al, gave 50% conversion (with respect to hydrogenation). All lithium and Grignard reagents displayed full conversion, although the nature of the catalyst obtained was tested only for EtMgCl, which had been chosen as a model substrate.

Regarding the solvent effect, it was observed that full solubility of substrates was not required for the quantitative conversion of norbornene. Among tetrahydrofuran, toluene, heptane, dioxane, ethyl acetate, and diethylether, only ethyl acetate and heptane did not give 100% yield. THF was indicated as the most appropriate solvent as it allowed for full solubility and redispersion of dried iron nanocatalyst.

Selectivity reported in the first attempt of Phua was later corroborated [46,62]. Terminal C=C bonds are more accessible and, thus can be hydrogenated in milder conditions, whereas internal bonds need higher temperature, pressure, or longer reaction times to reach comparable conversions. Kelsen at al. observed an effect of steric hindrance in the hydrogenation of internal alkenes [63]. While 4-octene was unreactive, 2-pentene exhibited a conversion as high as 95%. The two contrasting behaviors were explained by the increased steric hindrance of 4-octene, whose preferable configuration is transconfiguration, whereas 2-pentene dynamically changes configuration and, therefore, it is more accessible compared to 4-octene.

Generally, iron nanoparticles employed by Kelsen were somewhat more effective in comparison to results obtained via Grignard reduction by Phua. Substituted and nonsubstituted styrene, stilbene, and cycloalkenes were successfully hydrogenated under room temperature, with yields near 100%. The authors attribute this activity increase to the smaller size (~1.5 nm) and well-defined shapes of nanoparticles, which they obtained by the newly derived route. In this new synthesis route, $\{Fe(N[Si(CH_3)_3]_2)_2\}_2$ is decomposed under 3 bar of hydrogen at 150°C. It was mentioned that in the course of this reaction, the released $HN[Si(CH_3)_3]_2$ (HMDS) interacts with the iron nanocrystals. Another factor contributing to the activity of these nanoparticles is the absence of the oxide shell (that is usually formed), which was confirmed by a hydrides titration experiment. On the other hand, such small, reduced nanoparticles are extremely sensitive to air and water, which makes them a valuable and interesting model for further improvement and search for more stable catalysts.

Supported iron nanocatalysts exhibiting magnetic properties were proposed by Stein et al. [62]. Chemically derived graphene (CDG) was chosen as an easy-to-prepare resin, with high surface area. The supported nanocatalysts were obtained within 30 minutes by seeding with ultrasonic cavitation of iron pentacarbonyl on the presynthesized CDG sheets. The reaction was performed in diphenylmethane at room temperature. CDG synthesis comprised subsequent oxidation of graphite to graphene oxide and reduction/exfoliation. The composite materials, after separation from reaction media (via condensation of all by-products/solvent in high vacuum, at −196°C), were tested in catalytic hydrogenation of different alkenes. Presumably, due to some remaining oxidic functions on CDG, iron nanoclusters were partially oxidized on the surface. For this reason, to achieve full catalytic activity, addition of a small amount of reducing agent EtMgCl was required. Hydrogenation gave the corresponding alkanes with efficiency (and selectivity) similar to that reported by Phua et al. [61]. An important advantage of this material is that its magnetic character makes it easily isolable by magnetic decantation and therefore reusable. The unique magnetic properties, together with the heterogeneous nature of the composite, are prospective from the point of view of future studies and application.

The graphene/Fe composite materials were not described in terms of stability; however, former studies proved that iron catalysts are extremely prone to oxidation by both air and water. An important contribution, propelling the field toward more stable catalytic systems, was made by Hudson et al. [46]. Their strategy was based on simultaneous activation of iron and its protection by amphiphilic polymer support, which they used in aqueous flow hydrogenation. High conversions were achieved for a wide range of different compounds, including alkenes.

Hudson suggested two iron preparation pathways, which are depicted in Figure 4.7. The first synthetic route (*Scheme 1*) is widely applied thermal decomposition of iron carbonyl in the presence of polystyrene (PS) beads and an addition of linkers (LK). It proved to be more effective and gave higher conversions. The second route, incorporating tea extract as a reducing agent (*Scheme 2*), is, on the other hand, an environmentally benign option, which also provides iron nanocatalyst that is active in hydrogenation, although giving lower yields. The spacing in nanoparticles' crystal lattice was characteristic of bcc or fcc zero-valent iron, from which it had been assumed that there was no oxide shell on the nanoparticle surface. The loading of iron nanoparticles was determined by the choice of the linker ($-NH_2$ linker gave the highest loading of 11.72 mg/g)

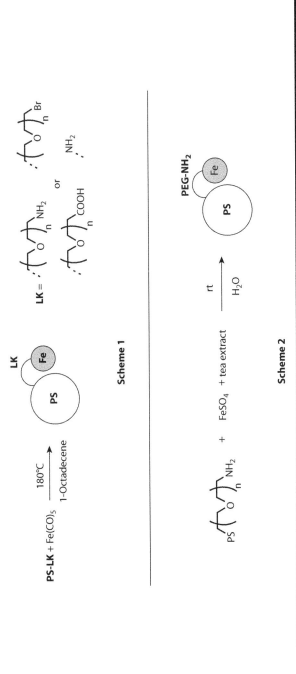

Figure 4.7 Two synthesis pathways for PS/iron catalyst proposed by Kelsen et al. (Reproduced from R. Hudson et al., *Green Chem.* 15 (2013) 2141. With permission.)

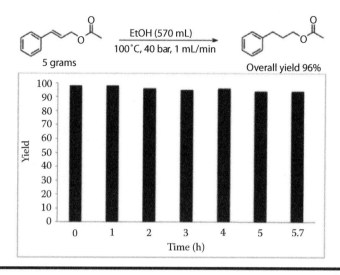

Figure 4.8 **Selective hydrogenation of cinnamyl acetate in ethanol scaled to large quantities. (Reproduced from R. Hudson et al., *Green Chem.* 15 (2013) 2141. With permission.)**

and the synthesis route (the tea extracts method, being less efficient, gave smaller loading). Hydrogenation reactions were carried out at 100°C and 40 bar of hydrogen, in ethanol, although for the model substrate, styrene, reaction tests utilizing water/ethanol mixtures and pure water were performed. Under elevated pressure conditions (60 bar), in ethanol/water mixtures of 50:50 and 10:90, ethyl benzene was produced with yields of 95% and 88%, respectively. Another experiment tested the possibility of scaling up the transformation for cinnamyl acetate (Figure 4.8). Only a minor decrease in reaction yield occurred, which could be a result of partial oxidation of iron nanoparticles or excessive packing of the catalyst.

4.3.2 Hydrogenation of Alkynes

All of the iron nanocatalysts discussed previously were also found active on alkyne hydrogenation, giving alkanes or alkenes. The results they obtained (summarized in Table 4.3)

Table 4.3 Alkyne Hydrogenation with Colloidal or Supported Iron Nanoparticles

Alkyne	Yield (%)	Selectivity (%)	Ref.
	100	100	[61]
	100	100	[61]
	67	100	[63]
	79	87	[63]
	7	100	[46]
	100	100	[61,64]
	0	—	[64]
	>99	100	[64]
	>89	100	[64]
	100	100	[46]
	90		[64]

(Continued)

Table 4.3 (Continued)   Alkyne Hydrogenation with Colloidal or Supported Iron Nanoparticles

Alkyne	Yield (%)	Selectivity (%)	Ref.
	12	12	[64]
	100	0 (Alkene)	[61]
	43	77	[61]

varied slightly for different catalytic systems; nevertheless, structure–activity relationships were noticed. For instance, similarly to alkenes, terminal aliphatic alkynes were more reactive compared to internal aliphatic substrates. However, different substituents can affect or even completely change this correlation. Generally, in all cases aryl substituents induce higher reactivity and usually full conversion of phenylacetylenes is realized [61,64].

This effect was especially visible in the case of graphene-supported iron nanoparticles. They exhibit negligible or no activity toward terminal (0% on 1-octyne conversion) and internal (7% on 4-octyne conversion) alkynes. However, diphenylacetylene afforded reasonable yields ~24%. The activity was dramatically enhanced due to the presence of a phenyl group in the terminal substrate, and quantitative hydrogenation for 1-phenyl-1-propyne was achieved [46]. On the other hand, the presence of a hydroxyl group retards formation of alkanes [61,64], and thus it offers some degree of selectivity toward alkenes. To explain this effect, Kelsen et al. [63] suggested that such compounds coordinate to iron nanoparticles via hydroxyl groups, which causes a decrease in the reaction rate. On the other hand, Hudson et al. [46] tested only aliphatic alkynes. Their polymer-supported catalyst gave comparable high conversions for both terminal and internal substrates.

Phua and co-workers offered some insight into the mechanism of alkyne hydrogenation [62]. An induction period for terminal and aliphatic alkynes was observed in which 1-octene and octane were produced at a fixed ratio. The reason for this seems to be due to nanoparticle aging, which changes the activity of the catalyst due to substrates or hydrogen binding. Basing on the observations and a Langmuir–Hinselwood model, they proposed the mechanism presented in Figure 4.9. The reaction proceeds via Fe-alkenyl complex (2), which rearranges to iron hydride alkylidene (3) and then can be hydrogenated and eventually form an alkane molecule.

Recently, significant progress has been made by Gieshoff et al. [60], who reported for the first time a stereoselective system for alkyne semihydrogenation, utilizing iron nanoparticles.

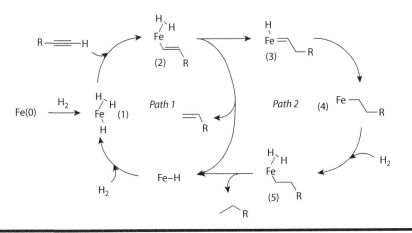

Figure 4.9 Proposed alkyne hydrogenation mechanism over iron catalysts. (Reproduced from M. Stein, J. Wieland, P. Steurer, F. Tölle, R. Mülhaupt, B. Breit, *Adv. Synth. Catal.* 353 (2011) 523. With permission.)

Their strategy incorporates ionic liquids based on azo-linium salts, thus forming a biphasic system that enabled multiple reuse/recycling. More importantly, they also developed promoter ligands that have an overwhelming impact on the selectivity of iron nanoparticles, leading to Z-selective semihydrogenation.

Gieshoff's nanocatalyst comprises three elements that were crucial for the whole nanocatalyst activity: (1) nanoparticles prepared via the Bedford-inspired method [35], (2) azolinium-based ionic liquid, and (3) species with nitrile or carbonyl moiety. Nanoparticles were ascribed as "low oxidation state" iron. They were approximately 4–5 nm in diameter, but increased during the reaction. Heptane appeared to be the most suitable solvent, since it allowed for the separation of the catalyst after the reaction. Ionic liquids were chosen as an alternative stabilizer, with low vapor pressure (good for stability). Hydrogenation reactions were performed in the presence of two ionic liquids: **IL1** and **IL2** (Figure 4.10), leading to strikingly different results. While **IL1**, containing no nitrile group in its structure, gave (almost exclusively) alkane as a product, nitrile-functionalized **IL2** displayed excellent selectivity to Z-alkene (95%).

To confirm whether it was the nitrile group that contributed to the observed selectivity, the influence of various additives was examined (Figure 4.11). The results showed that incorporating simple molecules with nitrile or carbonyl groups,

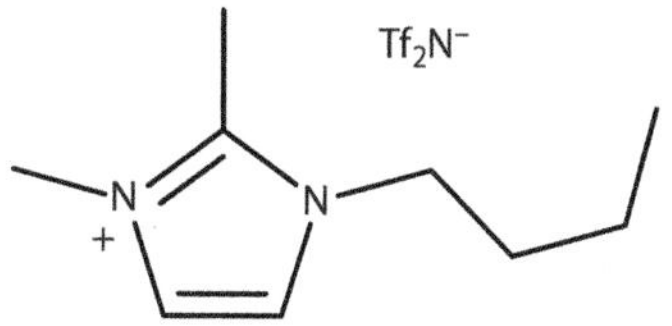

IL1 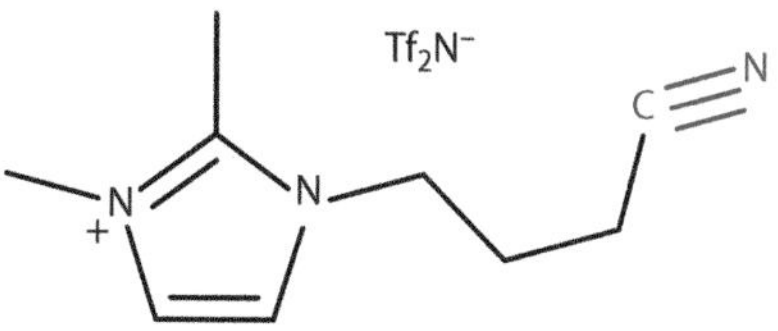   IL2

Figure 4.10   Two ionic liquids (IL) based on azolinium salts. (Reproduced from T. N. Gieshoff, A. Welther, M. T. Kessler, M. H. G. Prechtl, A. J. von Wangelin, *Chem. Commun.* 50 (2014) 2261. With permission.)

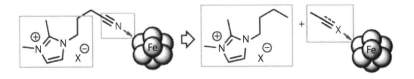

Figure 4.11 **Nucleophile-assisted hydrogenation with iron nano-particles, proposed by Geishoff et al. Two alternative scenarios: iron nanoparticles with IL2 or IL1+ nucleophilic molecule. (Reproduced from T. N. Gieshoff, A. Welther, M. T. Kessler, M. H. G. Prechtl, A. J. von Wangelin, *Chem. Commun.* 50 (2014) 2261. With permission.)**

together with **IL1**, gave a comparable or even the same reaction outcome, as it had been earlier observed with **IL2** alone. The selectivity-promoting effect is especially visible for acetonitrile. The addition of 50 to 200 mol% of MeCN favored an alkene product with 93% yield and with 96% Z-selectivity. Furthermore, employing the Fe/IL1/additive system significantly shortens the reaction time, from 2 days to 16 h. The procedure is also easier (not functionalized IL1). Eventually, this biphasic, nucleophile-assisted iron catalyst generated high yields of various alkynes (e.g., phenyl alkynes with NH_2, tBu, metoxy, or halide substituents), although terminal alkynes gave mixtures of partially and completely hydrogenated products.

4.3.3 Hydrogenation of C=O Bonds

Among the aforementioned iron nanocatalysts only the ones reported by Hudson et al. [46] show any activity toward carbonyl bond hydrogenation. Their polymer-supported iron nanoparticles allowed for high conversions of aromatic aldehydes (and imides), which were transformed into the corresponding alcohols (or amines) in aqueous flow conditions. The activating effect of the phenyl group was visible, as aliphatic aldehydes did not undergo the reaction. This was explained on the basis that aromatic compounds have a lower LUMO (lowest unoccupied molecular orbital) energy compared to aliphatic analogues. Ketones (being less accessible) were also not converted.

Sonneberg et al. reported a study of an asymmetric transfer hydrogenation of ketones [56]. Having previously worked on iron complex catalysts regarded as homogeneous, the authors raised a question of the true nature of their active species, which they suspected could be Fe(0) nanoparticles formed in the highly reducing reaction media. In order to provide evidence for this hypothesis, an array of analytical techniques were employed:

■ Kinetic and poisoning experiments to investigate the nature of catalytically active agents

■ STEM, energy dispersive x-ray spectroscopy (EDX) and XPS analysis to confirm the formation of zero-valent iron nanoparticles

■ Superconducting quantum interference device magnetometry (SQUID) to measure the magnetic properties of the catalyst

■ A combined poisoning/STEM/EDX experiment to confirm whether poisoning agents are binding to the Fe(0) surface

A model reaction transfer hydrogenation of acetophenone to 1-phenylethanol was conducted using isopropanol as a hydrogen source and KOtBu as a reducing agent, as shown in Figure 4.12.

Additionally, various alternative synthetic pathways were followed to test if the same result (including enantioselectivity) could be achieved when proceeding with different substrates, reducing agents, and reaction routes. These attempts involved incorporation of iron nanoparticles prepared via Grignard or carbothermal reduction, carbon supported iron nanoparticles, or iron(0) powders, with subsequent (or prior) addition of PNNP-ligands or precursors; however, none of these gave hydrogenated products.

The obtained results supported the hypothesis and thus it was concluded that 1-PNNP and 2-PNNP iron tetradentate

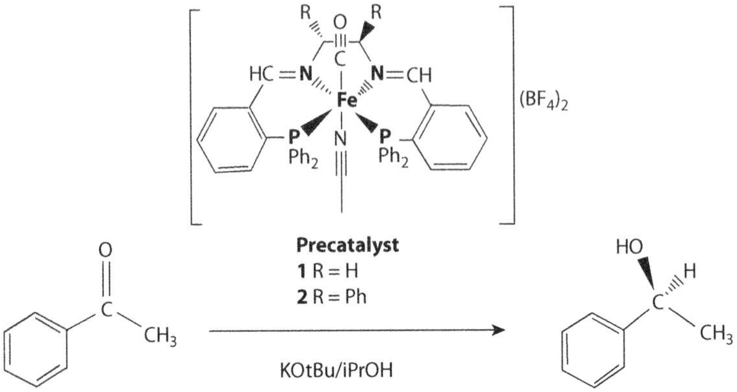

Figure 4.12   Scheme of asymmetric transfer-hydrogenation of acetophenone. Isopropanol was used as hydrogen source. (Reproduced from J. F. Sonnenberg, N. Coombs, P. A. Dube, R. H. Morris, *J. Am. Chem. Soc.* 134 (2012) 5893. With permission.)

complexes are indeed the catalyst precursors, which are reduced in the course of the reaction to give Fe(0) nanoparticles, which are stabilized by PNNP ligands (Figure 4.13). The coordination of chiral ligands is thought to determine the observed stereoselectivity. Although the catalyst structure has been well defined, the authors stress that it is very difficult to prove no influence of traces of homogeneous species on the reaction outcome.

Recently, Parimala and Santhanalakshmi [64] produced zerovalent iron nanoparticles immobilized on three different polymer resigns—namely, polyethylene glycol (PEG), carboxymethyl cellulose (CMC), or poly *N*-vinyl pyrrolidone (PVP) (Figure 4.14). Similar to the work by Hudson et al. [46], iron nanocatalysts were found to promote the reduction of substituted aromatic ketones to alcohols, however, instead of molecular hydrogen, they used $NaBH_4$ as a reducing agent. Substrates with different substituents in the aromatic ring were tested. Due to steric hindrance, *para*-substituents displayed higher reactivity in comparison to those in *ortho*-positions (Table 4.4). Also, electronreleasing groups were found to enhance the conversion, as opposed to electron withdrawing groups.

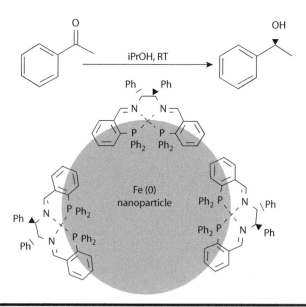

Figure 4.13 Heterogeneous iron(0) catalyst catalyzing asymmetric transfer hydrogenation of acetophenone. (Reproduced from J. F. Sonnenberg, N. Coombs, P. A. Dube, R. H. Morris, *J. Am. Chem. Soc.* 134 (2012) 5893. With permission.)

Figure 4.14 FESEM image of (a) Fe-PEG nanoparticles, (b) Fe-CMC nanoparticles, and (c) Fe-PVP nanoparticles. (Reproduced from L. Parimala, J. Santhanalakshmi, *J. Nanoparticles* 2014 (2014) 1. With permission.)

Table 4.4   Parameters of Reduction Reactions of Aromatic Ketones by NaBH$_4$ with Iron Nanoparticles as Catalysts

Entry	Aromatic Ketones	Product	Fe-PEG nps		Fe-CMC nps		Fe-PVP nps	
			Time (h)	% yield	Time (h)	% yield	Time (h)	% yield
1	acetophenone	1-phenylethanol	5	72	5	70	55	70
2	4-nitroacetophenone	1-(4-nitrophenyl)ethanol	6	68	6	65	6	65
3	2-nitroacetophenone	1-(2-nitrophenyl)ethanol	8	63	8.5	60	8.5	59
4	4-chloroacetophenone	1-(4-chlorophenyl)ethanol	4	75	4.5	73	4	71
5	4-bromoacetophenone	1-(4-bromophenyl)ethanol	3.5	77	4	75	4	73
6	4-methoxyacetophenone	1-(4-methoxyphenyl)ethanol	3.5	79	3.5	77	4	76
7	2-methoxyacetophenone	1-(2-methoxyphenyl)ethanol	6.5	65	6.5	63	7	61
8	4-methylacetophenone	1-(4-methylphenyl)ethanol	3	80	3	78	3	77
9	4-isobutylacetophenone	1-(4-isobutylphenyl)ethanol	3	85	3	80	3.5	77
10	2-chloro-1-phenylethanone	2-chloro-1-phenylethanol	5	70	5.5	68	5.5	67

Source: Reproduced from L. Parimala, J. Santhanalakshmi, *J. Nanoparticles* 2014 (2014) 1. With permission.

Note: Percentage yield attained during reaction time (h).

4.4 Concluding Remarks

The use of metallic iron nanoparticles in heterogeneous che-moselective hydrogenations is still scarce. However, this is forecast to change as the demand for cheap and selective hydrogenation catalysts increases, especially in the area of pharmaceuticals and fine chemicals. The current limitation hindering their expansion is, first, how to prepare small iron metallic nanoparticles from harmless sources and, second, to stabilize them so they can be used with greener solvents like water and in the presence of oxygen.

Mimicking nature will play a crucial role in the development of novel iron catalysts since nature hydrogenases containing iron active centers are the enzymes with the faster turnovers known to man. Finally, when possible, the processes should be carried out in flow, which often determines the type of support one can use in terms of nominal size. Polymers seem to be a great strategy to support and stabilize iron but their application is hindered by relatively low thermal stability. Thus, novel or modified polymer systems are required, possibly as composites.

References

1. S. Shima, R. K Thauer, A third type of hydrogenase catalyzing H_2 activation, *Chemical Record* (New York) 7 (2007) 37.
2. C. Madden, M. D. Vaughn, I. Díez-Pérez, K. A. Brown, P. W. King, D. Gust, A. L. Moore, T. A. Moore, Catalytic turnover of [FeFe]-hydrogenase based on single-molecule imaging, *J. Am. Chem. Soc.* 134 (2012) 1577.

3. W. Lubitz, H. Ogata, O. Rüdiger, E. Reijerse, Hydrogenases, *Chem. Rev.* 114 (2014) 4081.

4. D. L. Huber, Synthesis, properties, and applications of iron nanoparticles, *Small* 1 (2005) 482.

5. M. Melander, K. Laasonen, H. Jónsson, Effect of magnetic states on the reactivity of FCC(111) iron surface, *J. Phys. Chem. C* 118 (2014) 15863.

6. J. Sá, J. Szlachetko, M. Sikora, M. Kavčič, O. V. Safonova, M. Nachtegaal, Magnetic manipulation of molecules on a non-magnetic catalytic surface, *Nanoscale* 5 (2013) 8462.

7. C. Bolm, A new iron age, *Nat. Chem.* 1 (2009) 420.

8. M. Appl, Ammonia. In *Ullmann's encyclopedia of industrial chemistry*, Wiley-VCH, (2006).

9. References within: Paul J. Chirik, Modern alchemy: Replacing precious metals with iron in catalytic alkene and carbonyl hydrogenation reactions. In *Catalysis without precious metals*, R. M. Bullock (ed.), Wiley-VCH (2010).

10. Y. Hwang, Y.-C. Lee, P. D. Mines, Y. S. Huh, H. R. Andersen, Nanoscale zero-valent iron (nZVI) synthesis in a Mg-aminoclay solution exhibits increased stability and reactivity for reductive decontamination, *Appl. Catal. B* 147 (2014) 748.

11. X. Guan, Y. Sun, H. Qin, J. Li, I. M. C. Lo, D. He, H. Dong, The limitations of applying zero-valent iron technology in contaminants sequestration and the corresponding counter-measures: The development in zero-valent iron technology in the last two decades (1994–2014), *Water Res.* 75 (2015) 224.

12. A. Gual, C. Godard, S. Castillón, D. Curulla-Ferré, C. Claver, Colloidal Ru, Co and Fe-nanoparticles. Synthesis and application as nanocatalysts in the Fischer–Tropsch process, *Catal. Today* 183 (2012) 154.

13. Y. Koltypin, N. Perkas, A. Gedanken, Commercial edible oils as new solvents for ultrasonic synthesis of nanoparticles: The preparation of air stable nanocrystalline iron particles, *J. Mater. Chem.* 14 (2004) 2975.

14. H. Zhang, F. Liu, H. Jin, D. Liu, G. Que, Chemical synthesis of pure Fe nanoparticles under hydrogen atmosphere, *Chem. Bull./Huaxue Tongbao* 73 (2010) 377.

15. A.-H. Lu, E. L. Salabas, F. Schüth, Magnetic nanoparticles: Synthesis, protection, functionalization, and application, *Angew. Chem. Int. Ed.* 46 (2007) 1222.

16. T. W. Smith, D. Wychick, Colloidal iron dispersions prepared via the polymer-catalyzed decomposition of iron pentacarbonyl, *J. Phys. Chem.* 84 (1980) 1621.
17. L.-M. Lacroix, N. F. Huls, D. Ho, X. Sun, K. Cheng, S. Sun, Stable single-crystalline body centered cubic Fe nanoparticles, *Nano Lett.* 11 (2011) 1641.
18. K. S. Suslick, S.-B. Choe, A. A. Cichowlas, M. W. Grinstaff, Sonochemical synthesis of amorphous iron, *Nature* 353 (1991) 414.
19. K. S. Suslick, M. Fang, T. Hyeon, R. V. May, Sonochemical synthesis of iron colloids, *J. Am. Chem. Soc.* 118 (1996) 11960.
20. A. Tavakoli, M. Sohrabi, A. Kargari, A review of methods for synthesis of nanostructured metals with emphasis on iron compounds, *Chem. Papers* 61 (2007) 151.
21. J. H. Bang, K. S. Suslick, Applications of ultrasound to the synthesis of nanostructured materials, *Adv. Mater.* 22 (2010) 1039.
22. D. D. Caro, T. O. Ely, A. Mari, B. Chaudret, E. Snoeck, M. Respaud, J. Broto, Synthesis, characterization, and magnetic studies of nonagglomerated zerovalent iron particles. Unexpected size dependence of the structure, *Chem. Mater.* 8 (1996) 1987.
23. A. Gedanken, Using sonochemistry for the fabrication of nanomaterials, *Ultrasonics Sonochem.* 11 (2004) 47.
24. M. N. Nadagouda, A. B. Castle, R. C. Murdock, S. M. Hussain, R. S. Varma, *In vitro* biocompatibility of nanoscale zerovalent iron particles (NZVI) synthesized using tea polyphenols, *Green Chem.* 12 (2010) 114.
25. R. Herrera-Becerra, C. Zorrilla, J. L. Rius, J. A. Ascencio, Production of iron oxide nanoparticles by a biosynthesis method: An environmentally friendly route, *Appl. Phys. A: Mater. Sci. Process* 91 (2008) 241.
26. E. C. Njagi, H. Huang, L. Stafford, H. Genuino, H. M. Gallindo, J. B. Collins, G. E. Hoag, S. L. Suib, Biosynthesis of iron and silver nanoparticles at room temperature using aqueous sorghum bran extracts, *Langmuir* 27 (2011) 264.
27. A. Bharde, D. Rautaray, V. Bansal, A. Ahmad, I. Sarkar, S. M. Yusuf, M. Sanyal, M. Sastry, Extracellular biosynthesis of magnetite using fungi, *Small* 2 (2006) 135.
28. N. D. Meeks, V. Smuleac, C. Stevens, D. Bhattacharyya, Iron-based nanoparticles for toxic organic degradation: Silica platform and green synthesis, *Ind. Eng. Chem. Res.* 51 (2012) 9581.

29. D. P. E. Dickson, Nanostructured magnetism in living systems, *J. Magn. Magn. Mater.* 203 (1999) 46.

30. D. R. Lovley, J. F. Stolz, G. L. Nord, Jr., E. J. P. Phillips, Anaerobic magnetite production by a marine, magnetotactic bacterium, *Nature* 330 (1987) 252.

31. A. Bharde, A. Wani, Y. Shouche, P. A. Joy, B. L. V. Prasad, M. Sastry, Bacterial aerobic synthesis of nanocrystalline magnetite, *J. Am. Chem. Soc.* 127 (2005) 9326.

32. K. B. Narayanan, N. Sakthivel, Biological synthesis of metal nanoparticles by microbes, *Adv. Colloid Interface Sci.* 156 (2010) 1.

33. K. Huang, S. H. Ehrman, Synthesis of iron nanoparticles via chemical reduction with palladium ion seeds, *Langmuir* 34 (2007) 1419.

34. S.-N. Zhu, G. Liu, K. S. Hui, Z. Ye, K. N. Hui, A facile approach for the synthesis of stable amorphous nanoscale zero-valent iron particles, *Electronic Mater. Lett.* 10 (2014) 143.

35. R. B. Bedford, M. Betham, D. W. Bruce, S. A. Davis, R. M. Frost, M. Hird, Iron nanoparticles in the coupling of alkyl halides with aryl Grignard reagents, *Chem. Commun.* (2006) 1398–400.

36. S. Li, W. Yan, W. Zhang, Solvent-free production of nanoscale zero-valent iron (nZVI) with precision milling, *Green Chem.* 11 (2009) 1618.

37. http://www.nanoiron.cz.

38. R. Nakao, H. Rhee, Y. Uozumi, Hydrogenation and dehalogenation under aqueous conditions with an amphiphilic-polymer-supported nanopalladium catalyst, *Org. Lett.* 7 (2005) 2003.

39. D. Mahajan, P. Gu, Evaluation of nanosized iron in slurry-phase Fischer–Tropsch synthesis, *Energy & Fuels* 17 (2003) 1210.

40. D. Mahajan, P. Gütlich, U. Stumm, The role of nano-sized iron particles in slurry phase Fischer–Tropsch synthesis, *Catal. Commun.* 4 (2003) 101.

41. H. Schulz, Short history and present trends of Fischer–Tropsch synthesis, *Appl. Catal. A* 186 (1999) 3.

42. H. J. Schulte, B. Graf, W. Xia, M. Muhler, Nitrogen- and oxygen-functionalized multiwalled carbon nanotubes used as support in iron-catalyzed, high-temperature Fischer–Tropsch synthesis, *ChemCatChem* 4 (2012) 350.

43. N. van Thien Duc, S. Sufian, N. Mansor, N. Yahya, Investigation of carbon nanofiber supported iron catalyst preparation by deposition precipitation, *Adv. Mater. Res.* 1043 (2014) 71.

44. O. Margeat, F. Dumestre, C. Amiens, B. Chaudret, P. Lecante, M. Respaud, Synthesis of iron nanoparticles: Size effects, shape control and organization, *Progress Solid State Chem.* 33 (2005) 71.

45. M. Bhaumik, C. Noubactep, V. K. Gupta, R. I. McCrindle, A. Maity, Polyaniline/Fe^0 composite nanofibers: An excellent adsorbent for the removal of arsenic from aqueous solutions, *Chem. Eng. J.* 271 (2015) 135.

46. R. Hudson, G. Hamasaka, T. Osako, Y. M. A. Yamada, C.-J. Li, Y. Uozumi, A. Moores, Highly efficient iron(0) nanoparticle-catalyzed hydrogenation in water in flow, *Green Chem.* 15 (2013) 2141.

47. H. Naderpour, M. Noroozifar, M. Khorasani-Motlagh, Photodegradation of methyl orange catalyzed by nanoscale zerovalent iron particles supported on natural zeolite, *J. Iranian Chem. Soc.* 10 (2013) 471.

48. N. H. Khdary, M. A. Ghanem, M. G. Merajuddine, F. M. Bin Manie, Incorporation of Cu, Fe, Ag, and Au nanoparticles in mercapto-silica (MOS) and their CO_2 adsorption capacities, *J. CO_2 Utilization* 5 (2014) 17.

49. Z.-J. Li, L. Wang, L.-Y. Yuan, C.-L. Xiao, L. Mei, L.-R. Zheng, W.-Q. Shi, Efficient removal of uranium from aqueous solution by zero-valent iron nanoparticle and its graphene composite, *J. Hazard. Mater.* 290C (2015) 26.

50. X. Lv, X. Xue, G. Jiang, D. Wu, T. Sheng, H. Zhou, X. Xu, Nanoscale zero-valent iron (nZVI) assembled on magnetic Fe_3O_4/graphene for chromium (VI) removal from aqueous solution, *J. Coll. Interf. Sci.* 417 (2014) 51.

51. M. J. Asay, S. P. Fisher, S. E. Lee, F. S. Tham, D. Borchardt, V. Lavallo, Synthesis of unsymmetrical N-carboranyl NHCs: Directing effect of the carborane anion, *Chem. Commun.* 51 (2015) 5359.

52. F. He, M. Zhang, T. Qian, D. Zhao, Transport of carboxymethyl cellulose stabilized iron nanoparticles in porous media: Column experiments and modeling, *J. Coll. Interf. Sci.* 334 (2009) 96.

53. K. Yan, A. Chen, Selective hydrogenation of furfural and levulinic acid to biofuels on the ecofriendly Cu–Fe catalyst, *Fuel* 115 (2014) 101.

54. B. Zhu, T. Lim, Catalytic reduction of chlorobenzenes with Pd/Fe nanoparticles: Reactive sites, catalyst stability, particle aging, and regeneration, *Environ. Sci. Technol.* 41 (2007) 7523.

55. J.-M. Andanson, S. Marx, A. Baiker, Selective hydrogenation of cyclohexenone on iron–ruthenium nano-particles suspended in ionic liquids and CO_2-expanded ionic liquids, *Catal. Sci. Technol.* 2 (2012) 1403.

56. J. F. Sonnenberg, N. Coombs, P. A. Dube, R. H. Morris, Iron nanoparticles catalyzing the asymmetric transfer hydrogenation of ketones, *J. Am. Chem. Soc.* 134 (2012) 5893.

57. R. M. Bullock, *Catalysis without precious metals*, Wiley-VCH (2010) p. 302.

58. H.-U. Blaser, C. Malan, B. Pugin, F. Spindler, H. Steiner, M. Studer, selective hydrogenation for fine chemicals: Recent trends and new developments, *ChemInform* 34 (2003) 18.

59. J. G. De Vries, *The handbook of homogeneous hydrogenation*, Wiley-VCH (2005) p. 1595.

60. T. N. Gieshoff, A. Welther, M. T. Kessler, M. H. G. Prechtl, A. J. von Wangelin, Stereoselective iron-catalyzed alkyne hydrogenation in ionic liquids, *Chem. Commun.* 50 (2014) 2261.

61. P.-H. Phua, L. Lefort, J. A. F. Boogers, M. Tristany, J. G. de Vries, Soluble iron nanoparticles as cheap and environmentally benign alkene and alkyne hydrogenation catalysts, *Chem. Commun.* 25 (2009) 3747.

62. M. Stein, J. Wieland, P. Steurer, F. Tölle, R. Mülhaupt, B. Breit, Iron nanoparticles supported on chemically-derived graphene: Catalytic hydrogenation with magnetic catalyst separation, *Adv. Synth. Catal.* 353 (2011) 523.

63. V. Kelsen, B. Wendt, S. Werkmeister, K. Junge, M. Beller, B. Chaudret, The use of ultrasmall iron(0) nanoparticles as catalysts for the selective hydrogenation of unsaturated C-C bonds, *Chem. Commun.* 49 (2013) 3416.

64. L. Parimala, J. Santhanalakshmi, Studies on the iron nanoparticles catalyzed reduction of substituted aromatic ketones to alcohols, *J. Nanoparticles* 2014 (2014) 1.

Chapter 5

Chapter 5

Hydrogenation by Silver Catalysts

Cristina Paun and Jacinto Sá

Contents

This chapter deals with hydrogenation mediated by silver metal present as supported and unsupported nanoparticles. In the chapter, we highlight synthetic procedures for the preparation of metallic silver nanoparticles and their applications in catalysis. The emphasis is on catalytic reactions for production of fine and pharmaceutical chemicals.

5.1 Introduction

Silver (Latin: *argentum*) is a soft, white, and lustrous transition metal that possesses the highest electrical conductivity of any element and the highest thermal conductivity and reflectivity of any metal. The metal occurs naturally in its pure, free form (native silver), as an alloy with gold and other metals, and in minerals such as argentite. Most silver is produced as a by-product of copper, gold, lead, and zinc refining. Silver has long been valued as a precious metal. More abundant than gold, silver metal has in many premodern monetary systems functioned as coinable specie, sometimes even alongside gold. In addition, silver has numerous applications beyond currency, such as in solar panels, water filtration, jewelry and ornaments, and high-value tableware and utensils (hence the term "silverware"). Silver is used industrially in electrical contacts and conductors, in specialized mirrors, and in window coatings. Its compounds are used in photographic film and x-rays. Dilute silver nitrate solutions and other silver compounds are used as disinfectants and microbiocides (oligodynamic effect) and added to bandages and wound dressings, catheters, and other medical instruments.

Silver also possesses catalytic properties ideal for use as an oxidation catalyst, for example, in the production of formaldehyde from methanol and air by means of silver crystallites containing a minimum 99.95 wt% silver. Supported silver is probably the only catalyst available today capable of converting ethylene to ethylene oxide (later hydrolyzed to ethylene glycol, used for the manufacturing of polyesters)—an important industrial reaction. It is also used in the Oddy test [1] to detect reduced sulfur compounds and carbonyl sulfides.

Silver is the metal with the lowest hydrogen-binding energy, confirmed by experimental results and theoretical calculations [2–4]. Thus, it is rather surprising that it can catalyze hydrogenation reactions with molecular hydrogen. Its activity is strictly related to its size and shape, which need to be in the nanoscale. This chapter about hydrogenation reactions catalyzed by silver is focused in systems in which silver is clearly identifiable as the active center, whereas systems where silver is presented as dopant, co-catalyst, or promoter are only marginally mentioned. The most notorious case is the Ag promotion of Pd catalysts used in tail-end acetylene hydrogenation. Readers are advised to consult the numerous publications and reviews on the topic published elsewhere [5–8].

5.2 Types of Hydrogenations

It is known that a catalytic reaction performs best if the interaction between the adsorbates and the surface is not too strong and not too weak. In the case of metal catalysts, this relates to the d-band occupancy. In this respect, hydrogenation on silver surface raises the question of whether H_2 proceeds in the mechanism of the reaction as molecule or dissociates. Silver has an electronic configuration, $4d^{10}5s^1$, and cannot dissociate the strong bond of H_2. However, based on the experience of gold (i.e., Au particles of nanosize dimensions can dissociatively chemisorb small molecules, e.g., O_2), small clusters of silver atoms have also been proposed as being capable of H_2 dissociation, therefore making silver catalysts as potentially active for certain hydrogenation reactions. The reason relates to a shift of electron density from 4d to the 5s, when you reduce the cluster size similar to what happens with gold nanoparticles [9,10]. In the following, the reader will find examples of such reactions, some with industrial relevance, together with studies of density functional theory (DFT) and different characterization techniques used to elucidate the reaction mechanism on small particles of silver.

5.2.1 Aldehyde Hydrogenation

The selective hydrogenation of α,β-unsaturated aldehydes to the corresponding allylic alcohols is an important reaction for the industrial production of fine chemicals as well as for fundamental research in catalysis. In the presence of conventional hydrogenation catalysts (e.g., supported Pt, Pd, Cu), α,β-unsaturated aldehydes are hydrogenated predominantly to the saturated aldehydes or even to saturated alcohols (Figure 5.1). Therefore, it is desirable to find catalysts that will control the intramolecular selectivity by hydrogenation, preferably the C=O group, while keeping the olefinic bond intact.

The use of Ag catalysts for the selective hydrogenation of α,β-unsaturated aldehydes was first proposed by Nagase et al. [11,12] in the late 1980s. They used a silver-manganese oxide catalyst in the selective hydrogenation of crotonaldehyde to crotyl alcohol. When the reaction was performed above 150°C, they were able to convert more than 85% of the substrate with 70% selectivity to the corresponding alcohol. The selectivity to

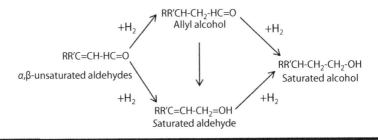

Figure 5.1   Reaction scheme for α,β-unsaturated aldehydes hydrogenation.

the unsaturated alcohol was found to be due to the low inter-action of the C=C group with the silver surface [13].

Claus et al. [14,15] showed that the hydrogenation of crotonaldehyde could be carried out over Ag supported on SiO_2 (Figure 5.2) and Al_2O_3, with selectivities similar to the Ag-MnO_2 catalysts. By means of electron microscopy, they suggested that the reaction is structure sensitive after obtaining higher turnover numbers and selectivities with catalysts containing mainly silver particles with dense Ag(111) surface planes; the dominating high-index planes in the smaller particles were found to be less selective (Figure 5.3) [16].

The result was found valid for the hydrogenation of acrolein and propanal [17]. Apart from the structure, the authors demonstrated that the selectivity depends on both hydrogen and aldehyde pressure, at least when the reaction is performed in the gas phase. Acrolein partial pressure affects the adsorption of the molecule, whereas hydrogen partial pressure controls hydrogen availability [18–22].

Bron et al. [23,24] measured a significant enhancement in the conversion level and, to a smaller extent, in the selectivity when the Ag/SiO_2 catalyst was preconditioned in oxygen. The measured effect led to the speculation that an optimum charge of the metal exists, and that positively charged silver promotes conversion but not selectivity.

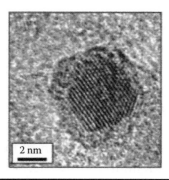

Figure 5.2 **Cuboctahedral Ag particle of catalyst Ag/SiO_2 covered by an amorphous shell. (Reproduced from P. Claus, H. Hofmeister, *J. Phys. Chem. B* 103 (1999) 2766. With permission.)**

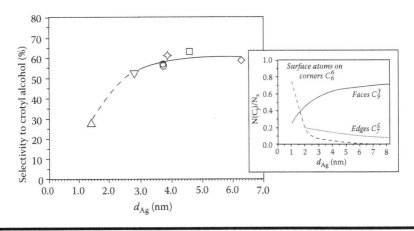

Figure 5.3 **Selectivity toward crotyl alcohol depending on the silver particle size. (– – –) Titania-supported silver catalysts: (△) Ag/TiO$_2$-high-temperature reduction, (▽) Ag/TiO$_2$ low-temperature reduction; (——) silica-supported silver catalysts: (○) Ag/SiO$_2$_sol-gel_1, (○) Ag/SiO$_2$_sol-gel_3, (◇) Ag/SiO$_2$_impregnation, (□) Ag/SiO$_2$_sol-gel_2, (◇) Ag/SiO$_2$_precipitation-deposition).** *Insert:* **Relative occurrence (N(C$_j$)/N$_s$) of various types of surface atoms (Cj) as a function of the silver particle size d$_{Ag}$ (according to van Hardeveld and Hartog [25]): C$_6^6$ = corner atom sites; C$_7^5$ = edge atom sites between cube and octahedron faces C$_9^3$ = face atoms on close-packed octahedron faces. (Reproduced from P. Claus, H. Hofmeister,** *J. Phys. Chem. B* **103 (1999) 2766. With permission.)**

The overall performance of the silver catalysts in the hydrogenation of α,β-unsaturated aldehydes is also dependent on support acid–base properties: Strong Lewis acid sites were found detrimental for conversion and selectivity to allyl alcohol. The authors reached the conclusion after measuring an improvement in the catalytic performance when the alumina content in the Al$_2$O$_3$/SiO$_2$ support was increased [26]. Furthermore, when supported on ZnO, silver was found less active and selective than Cu in the hydrogenation of but-2-enal [27].

The addition of indium to the Ag/SiO$_2$ catalysts led to a fourfold increase in the catalytic performance in the selective hydrogenation of acrolein [28] and citral [29]. Indium, which is positively charged during reaction, enhances acrolein coverage and strengthens acrolein–silver bonding, thus influencing the catalytic performance [30]. The conclusion was based on x-ray absorption near-edge spectroscopy (XANES) at In K-spectra (Figure 5.4), which showed clear differences depending on the reaction conditions. First of all, under reaction conditions as well as in the as-prepared catalyst, the indium is not in a pure metallic state. The hydrogenation catalyst must be a hetero-geneous mixture that contains oxidized indium compounds in addition to metallic silver and/or silver–indium particles on the SiO_2 carrier. Even under strongly reducing conditions, the indium remained oxidized. The subsequent addition of acrolein led to the formation of even more oxidized indium species in the catalyst material. The status of the working catalyst lies between the two extreme conditions of a reducing

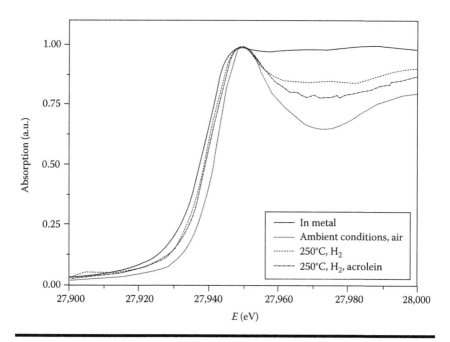

Figure 5.4   In K-XANES of 9Ag1.25In/SiO₂, recorded under different experimental conditions. Spectra are background corrected and normal-ized to the white line intensities. For better comparison purposes, a spec-trum of an In metal foil is given as well. (Reproduced from F. Haass, M. Bron, H. Fuess, P. Claus, *Appl. Catal. A* 318 (2007) 9. With permission.)

atmosphere (H_2) and a more oxidizing atmosphere (air), which is elucidated clearly by this XANES analysis.

Brandt et al. [31] demonstrated that acrolein coverage depends on the molecule adsorption geometry and is critical in determining selectivity. At low coverages and in the absence of adsorbed hydrogen, both the carbonyl and the alkene functionalities are almost parallel to the surface. At high coverages the C=C bond tilts markedly with respect to the surface, rendering it less vulnerable toward reaction with hydrogen adatoms, kinetically favoring the formation of the otherwise thermodynamically disfavored unsaturated alcohol.

Khanra, Jugnet, and Bertolini [32] suggested three possible adsorption geometries of acrolein on the surface of transition metals (Figure 5.5). DFT suggested that only di-σ_{CC} and η_3-*cis* are stable. Furthermore, the authors concluded that when the acrolein is adsorbed in di-σ_{CC}, the main product is propanal, whereas when acrolein adsorbs in η_3-*cis* geometry the allyl alcohol is formed. Figure 5.6 shows the reaction energies for acrolein hydrogenation.

DFT calculations done by Lim et al. [33] yielded information on acrolein adsorption geometry on Ag surfaces (Figures 5.7 through 5.9) and suggested that the hydrogenation of acrolein on Ag(110) yields propanal as the main product (Figures 5.10 and 5.11). The O_{sub}/Ag(111) model system was found less active with respect to acrolein hydrogenation by adsorbed H atoms

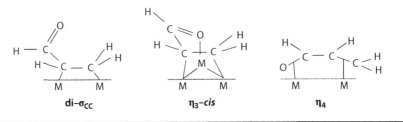

Figure 5.5 **Schematic geometries of adsorption of acrolein in di-σ_{CC}, η_3-*cis* and η_4 coordinations. (Reproduced from K. Brandt, M. E. Chui, D. J. Watson, M. S. Tikhov, R. M. Lambert, *J. Am. Chem. Soc.* 131 (2009) 17286. With permission.)**

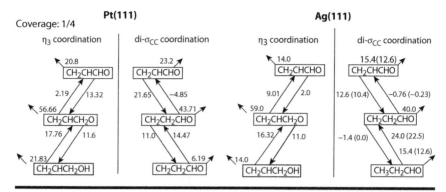

Figure 5.6 Reaction energetic for acrolein hydrogenation on Pt(111) and Ag(111) surface (all energies in kcals/mol). (Reproduced from B. C. Khanra, Y. Jugnet, J. C. Bertolini, *J. Mol. Catal. A* **208** (2004) 167. With permission.)

than the clean Ag surface. However, selectivity to allyl alcohol was found to be extremely high over O_{sub}/Ag(111). Contrary to other transition metal, such as Pt, the hydrogenation products desorb easily from both Ag(110) and O_{sub}/Ag(111) at common reaction temperatures.

Yang et al. [34] also performed DFT calculations and suggested that silver's high selectivity in the hydrogenation of crotonaldehyde is due to the preference of the C=O atop adsorption mode for the low-coordination sites located at the edges and corners of silver particles. This reveals a distinctive size effect on the hydrogenation of α,β-unsaturated aldehydes by silver.

5.2.2 Nitroaromatic Hydrogenation

Aniline and its derivatives are important intermediates for fine chemicals such as agrochemicals, pharmaceuticals, and pigments. They are mainly produced by the selective hydrogenation of the corresponding nitroaromatic. Chen et al. [35] demonstrated that Ag/SiO$_2$ could selectively convert chloronitrobenzene to chloroaniline at 140°C and a pressure higher than 5 bar of hydrogen. At 20 bar of hydrogen, 100%

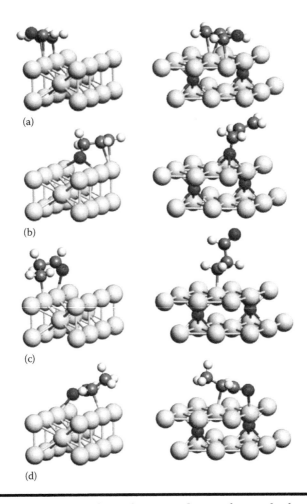

Figure 5.7 Calculated adsorption complexes of monohydrated intermediates of acrolein hydrogenation on the surfaces Ag(110) (left-hand column) and O_{sub}/Ag(111) (right-hand column): (a) hydroxyallyl (mh0); (b) allyloxy (mh1); (c) 2-formylethyl (mh2); and (d) 1-formylethyl (mh3). Only the top three layers of Ag(110) slab and two layers of O_{sub}/Ag(111) are shown. (Reproduced from K. H. Lim, A. B. Mohammad, I. V. Yudanov, K. M. Neyman, M. Bron, P. Claus, N. Rösch, *J. Phys. Chem. C* 113 (2009) 13231. With permission.)

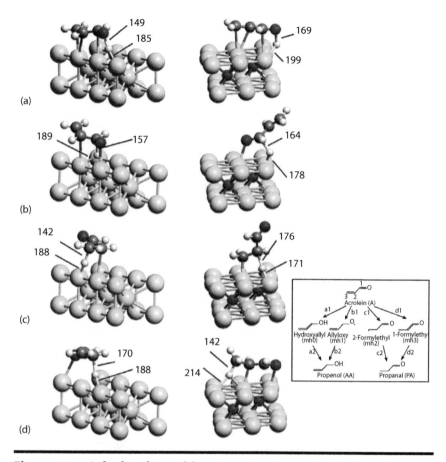

Figure 5.8 Calculated transition state structures for the first steps a1–d1 (insert) of the partial acrolein hydrogenation on the surfaces Ag(110) (left-hand column) and O_{sub}/Ag(111) (right-hand column). Also given are distances (in picometers) from the attacking atom H to the nearest atom of the reactant (H-X) and to the substrate (H-Ag). *Insert:* Elementary steps of the partial hydrogenation of adsorbed acrolein by atomic H studied on silver catalysts. (Reproduced from K. H. Lim, A. B. Mohammad, I. V. Yudanov, K. M. Neyman, M. Bron, P. Claus, N. Rösch, *J. Phys. Chem. C* 113 (2009) 13231. With permission.)

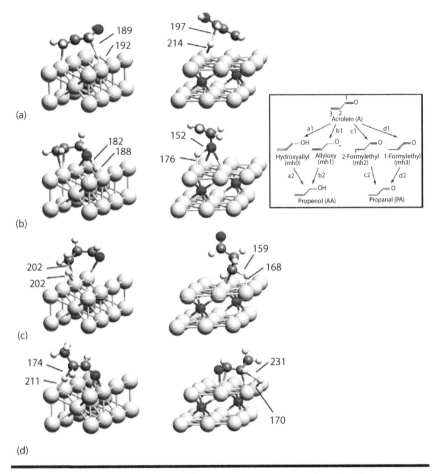

Figure 5.9 Calculated transition state structures for the first steps a1–d1 (insert) of the partial acrolein hydrogenation on the surfaces Ag(110) (left-hand column) and O_{sub}/Ag(111) (right-hand column). Also given are distances (in picometers) from the attacking atom H to the nearest atom of the reactant (H-X) and to the substrate (H-Ag). *Insert:* Elementary steps of the partial hydrogenation of adsorbed acrolein by atomic H studied on silver catalysts. (Reproduced from K. H. Lim, A. B. Mohammad, I. V. Yudanov, K. M. Neyman, M. Bron, P. Claus, N. Rösch, *J. Phys. Chem.* C 113 (2009) 13231. With permission.)

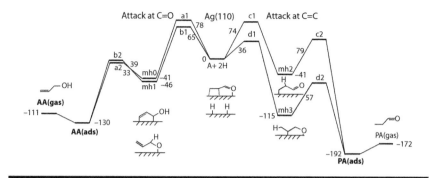

Figure 5.10 Reaction profile of the partial hydrogenation of acrolein (A) to propenol (AA) and propanal (PA) on Ag(110). Reaction and activation energies in kJ mol^{-1}. (Reproduced from K. H. Lim, A. B. Mohammad, I. V. Yudanov, K. M. Neyman, M. Bron, P. Claus, N. Rösch, *J. Phys. Chem. C* 113 (2009) 13231. With permission.)

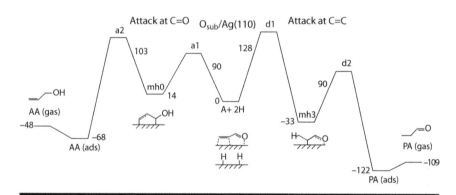

Figure 5.11 Reaction profile for the partial hydrogenation of acrolein (A) to propenol (AA) and propanal (PA) on the model surface O$_{sub}$/Ag(111). Pathways b and c are not shown as they exhibit significantly higher barriers than pathways a and d. Reaction and activation energies in kJ mol^{-1}. (Reproduced from K. H. Lim, A. B. Mohammad, I. V. Yudanov, K. M. Neyman, M. Bron, P. Claus, N. Rösch, *J. Phys. Chem. C* 113 (2009) 13231. With permission.)

selectivity and conversion was obtained. Similar results were obtained by Li, Li, and Xu [36] with Ag supported on C$_{60}$. Zhang, Li, and Chen [37] successfully tested Ag nanoclusters produced with the help of an ionic liquid in the hydrogenation of 4-nitrophenol with NaBH$_4$.

Recently, Shimizu, Miyamoto, and Satsuma [38] reported that nitroaromatic hydrogenation is structure sensitive after finding that silver clusters on θ-Al$_2$O$_3$ were highly chemoselective to the reduction of substituted nitroaromatics. The unsaturated Ag sites on the silver cluster were found responsible for the rate-limiting H$_2$ dissociation to yield a H$^+$/H$^-$ pair at the metal/support interface, while the basic site on Al$_2$O$_3$ acts as an adsorption site for nitroaromatics. The high chemoselectivity can be attributed to a preferential transfer of H$^+$/H$^-$ to the polar bonds in the nitro group, similar to what happens with Au/Al$_2$O$_3$ catalysts [39,40]. However, despite the high selectivities and conversions obtained with Ag and Au, the catalysts were significantly less active than Pt.

Leelavathi, Rao, and Pradeep [41] used quantum clusters (QCs) of Ag$_{7,8}$ supported on different oxides for the reduction reaction (with NaBH$_4$) of various aromatic nitro compounds. For Al$_2$O$_3$@Ag$_{7,8}$ (10% loading), the reduction reaction occurred with a rate constant of 8.23×10^{-3} s^{-1} at 35°C, and the TOF measured was 1.87 s^{-1} per cluster. In comparison with Ag nanoparticles (prepared by the citrate method) loaded on alumina, these new cluster catalysts were more active for the investigated reactions. Furthermore, the supported clusters remained active upon recycling.

5.2.3 Olefin Hydrogenation

As mentioned, silver catalysts are known to be effective in the selective hydrogenation of the C=O group of α,β-unsaturated aldehydes and –NO$_2$ group of nitroaromatics, which is a consequence of the oxophilicity of silver and the weak adsorption of unsaturated hydrocarbons. Therefore, it is not surprising that only very modest conversion in olefin hydrogenation are measured when Ag is used on its own [42], in contrast to copper and palladium catalysts, which are renowned olefin hydrogenation catalysts. Generally, to improve the catalytic performance, Ag catalysts are doped, for example, with lanthanides such as Eu and Yb [43]. Ehwald, Shestov, and Muzykantov [44] suggested that to be active, silver must be highly dispersed and present as small particles. However, no dedicated study on the influence of shape and size was performed to establish a definite connection between size, surface structure, and reactivity.

Sárkány and Révay [45] used Ag supported on TiO_2 and SiO_2 in the hydrogenation of acetylene, propene, 1-butene, and 1,3-butadiene. They observed lower rates for the hydrogenation of propene and 1-butene in comparison with acetylene and 1,3-butadiene, which they ascribed to the rather weak complexation of alkenes on Ag sites and not the inability of Ag to dissociate hydrogen. Silver dissociates hydrogen via heterolytic fission similar to Cu [46] and Au [9,47]. Mohammad et al. [48] calculated the activation of H_2 on clean planar (111), (110), and stepped (221) as well as oxygen precovered silver surfaces. They determined that clean silver is inert toward H_2 dissociation, both thermodynamically and kinetically. The reaction is endothermic by ~40 kJ mol^{-1} and exhibits high activation energies of ~125 kJ mol^{-1}. However, in the presence of oxygen on the surface, H_2 dissociation is exothermic and kinetically feasible. The reaction proceeds in two steps: First, the H–H bond is broken at an Ag–O pair with an activation barrier E_a ~70 kJ mol^{-1}; then the H atom bound at an Ag center migrates to a neighboring O center with E_a ~12 kJ mol^{-1}. In supported Ag catalysts, the support plays a role in the splitting of H_2.

The weak strength of adsorption of alkenes over Ag is a direct consequence of the lack of vacant d-orbitals. The olefilic C–C bond of adsorbed ethene on Ag is 1.88 Å [49], which suggests only slight distortion of the surface species. Interaction between double occupied π-orbital of ethene and filled surface d_z^2 is repulsive and the bonding fragment of surface d_{xy} and p_x and empty asymmetric π* orbitals leads to only weak complexation. In the case of 1,3-butadiene hydrogenation, Sárkány and Révay [45] detected high selectivity to 1-butene, which they attributed to the high preference of 1,2-interaction and probably to the formation of σ-butenyl intermediates in the half-hydrogenated state. 2-Butenes are likely to be formed via 1,4 addition of hydrogen and in the reaction πσ-bonded half-hydrogenated states or π-allyl states may participate. Moyes et al. [50] assigned the high selectivity to 1-butene to electronic effects on the mode of complexation of diene (Figures 5.12 and 5.13) and its half-hydrogenated state rather than the atomic distances suggested by Nishimura, Inoue, and Yasumori [51]. As with Au, the high selectivity of Ag in the semihydrogenation reaction is thus ascribed to the low rate of alkene hydrogenation.

The proposal of Ehwald et al. [44] that smaller silver particles are preferred to achieve better performances was not extensively studied—and even less the reactivity of silver with different exposed facets. The increase of catalytic performance with the decrease of particle size suggests that silver loses its

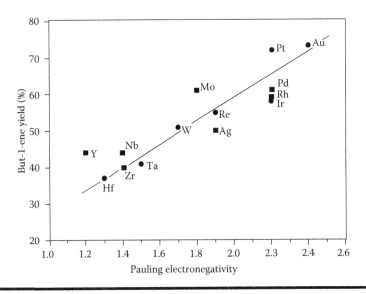

Figure 5.12 Correlation of but-1-ene yield with Pauling electronegativity for the elements of the second transition series (■) and the third transition series (●). (Reproduced from R. B. Moyes, P. B. Wells, J. Grant, N. Y. Salman, *Appl. Catal. A* 229 (2002) 251. With permission.)

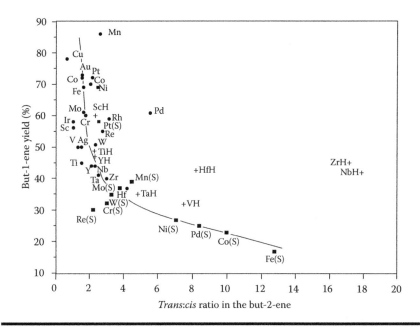

Figure 5.13 Interrelation between the but-1-ene yield and the *trans:cis* ratio in the but-2-ene yield. (●), Pure metals; (■), sulfided metals; (+), metal hydrides. (Reproduced from R. B. Moyes, P. B. Wells, J. Grant, N. Y. Salman, *Appl. Catal. A* 229 (2002) 251. With permission.)

metallic character, which might improve hydrogen activation. Somorjai and Aliaga [52] suggested that metal nanoparticles of less than 2 nm in diameter easily undergo changes of oxidation state, indicating changes of electronic structure (such as ionization potential and electron affinity). Several authors have confirmed this in the case for Pt and Au supported catalysts [53,54]. This is yet to be confirmed in the case of silver supported catalysts. Another issue that should be considered in the catalysis with Ag is the carbide formation during hydrogenation. It has been observed with other metals [55–59]; however, it has not been studied for silver catalysts.

Ag/SiO$_2$ was also found to be able to hydrogenate linoleic acid to oleic acid and stearic acid, with selectivities around 60% at low conversion [60]. At high conversion, the selectivity drops to roughly 20%. It must be emphasized that the aim was to isomerize linoleic acid to other conjugated linoleic acids.

5.2.4 CO_2/CO Hydrogenation (Methanol Synthesis)

Methanol synthesis is a process of major importance for the chemical industry. Methanol is synthesized from synthesis gas (syngas), which comprises a mixture of carbon monoxide, hydrogen, and carbon dioxide, over a copper zinc oxide supported on alumina, a process implemented by Imperial Chemical Industries in 1966. At 50–100 bar and 250°C, it can catalyze the production of methanol from carbon monoxide and hydrogen with a selectivity close to 100%. Apart from the common use as a laboratory solvent, methanol's main usage is in the production of other chemicals. About 40% of methanol is converted to formaldehyde and from there into products as diverse as plastics, plywood, paints, explosives, and permanent-press textiles.

Ag/SiO_2 was found active in the hydrogenation of CO_2 to methanol by Sugawa et al. [61]. The catalyst was found 33 times less active than its Cu counterpart and 345 times less than the most active Ru catalyst. However, the catalyst only yields methanol and CO as products. The activity increased 10-fold with the addition of ZnO, and the selectivity to methanol increased from 11.6% to 83.9%. Once again, the other product is CO and, most importantly, methane was not detected. Wambach, Baiker, and Wokaum [62] reported an increase in the methanol yield when Ag is added to a Cu/ZrO_2. A similar result was obtained with Ag-promoted Cu/$CrAl_3O_6$ [63], whereas Vilchis-Nestor et al. [64] reported a decrease of the yield of methane production in CO hydrogenation when Ag was added to their Au/Al_2O_3/SiO_2. Shaw et al. [65] studied methanol synthesis over Ag/CeO_2 and Cu,Ag/CeO_2 and concluded that the metal phase component is crucial for the production of methanol; they ruled out that the chemistry occurs solely on the oxide phase, despite finding dominant interactions between Ag and CeO_2.

5.2.5 Chemoselective Alkyne Hydrogenation

There are no published studies on alkyne chemoselective hydrogenation. We prepared silver nanoparticles with well-defined shapes (cubes and near spheres) and sizes (45 nm, 17 nm) with the polyol process [66]. Ethylene glycol was used as both solvent and reducing agent. The control of the shape of the two silver colloids has been attained by varying the ratio of poly (vinyl pyrrolidone) (PVP) to Ag, and by adding an etching agent ($Na_2S \cdot 9H_2O$) in the case of the nanocubes.

Characterization of the nanoparticles with transmission electron microscopy (TEM) and high-resolution transmission electron microscopy (HRTEM) (small inlets) in Figure 5.12 revealed that the cubes had mainly {100} exposed facets, whereas the spheres had mainly {111} exposed facets—making them the perfect candidates for study of the effect of the exposed facet on catalyst reactivity. Figure 5.14a and b shows the TEM images of the prepared colloids after washing; the images are representative of the materials prepared. The particle size distributions of the colloids were measured from the TEM observations. Figure 5.14c and d shows the histograms corresponding to the particle size distributions of the prepared colloids based on a 100-particle count. The average particle sizes of the prepared colloids, as measured by TEM, were 45 nm for cubes and 17 nm for spheres. The results indicated that the preparation methods allowed us to obtain particles with well-defined shapes with a narrow size distribution (monodisperse). A measure of monodispersity is the variation coefficient, which should be below 20% in order to have a monodisperse colloid. The variation coefficient for the cubes was 11%, and 17% for the spheres.

These silver nanoparticles were immobilized on three different supports (γ-Al_2O_3 155 m²/g, SiO_2 490 m²/g, TiO_2 35–65 m²/g). For comparison, we also prepared a series of catalysts by the classical wet impregnation from $AgNO_3$. Figure 5.15 presents the TEM images of the catalysts obtained by the two methods.

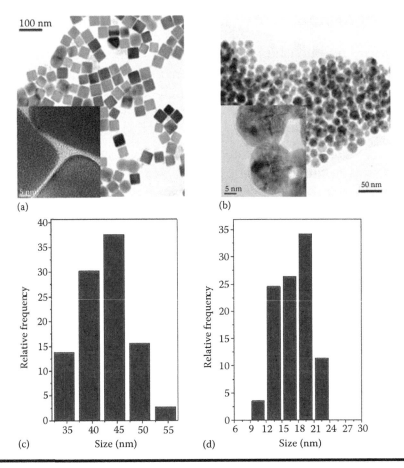

Figure 5.14 TEM/HRTEM (inlets) of (a) nanocubes and (b) nanospheres, and the particle size distribution of the (c) nanocubes and (d) nanospheres.

It is evident from the images that after the deposition on the supports, the shape and size of the cube and sphere particles were well maintained (particularly well for the cubes), whereas the materials obtained by impregnation (Ag/TiO_2 and Ag/SiO_2) had a large particle size distribution (0.7–25 nm in a single sample). Therefore, the benefit of preparing heterogeneous materials starting from colloids with well-defined shapes and sizes can be clearly observed.

The Ag colloids (cubes, spheres), the supported well-defined nanoparticles, and the impregnated catalysts, as well

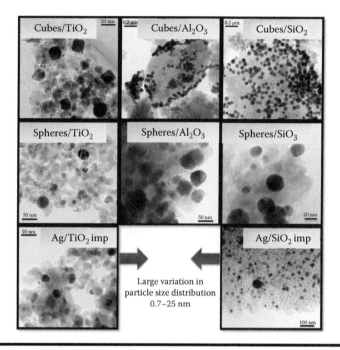

Figure 5.15 TEM of supported silver catalysts.

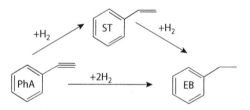

Scheme 5.1 Phenylacetylene (PhA) hydrogenation to styrene (ST) and ethylbenzene (EB).

as the bare supports, have been tested in phenylacetylene hydrogenation (Scheme 5.1). The reactions were performed in two systems in gas and liquid phase.

5.2.5.1 Gas-Phase Hydrogenation of Phenylacetylene

The catalytic hydrogenation of phenylacetylene (PhA) was carried out in a plug flow reactor at ambient pressure. The reactor was placed in an electric oven at 300°C. The PhA was introduced in the system by flowing the gas mixture (7.7% H_2/He, Ar) through a saturator that contained the former, with the molar ratio PhA:H_2 = 1:8. The analysis of the reaction's products was done with a mass spectrometer coupled online. Prior to reaction, the catalysts were reduced in situ at 300°C for 1 hour, to remove any traces of stabilizing agent. The summarized results are presented in Table 5.1.

The data revealed that the conversions of PhA were very modest, with the best catalysts from the series tested being the impregnated materials. In terms of selectivity, the cubes supported on the three supports seemed to give higher selectivities to styrene over the three runs, which would suggest a positive influence of the exposed {100} facet. However, on a closer examination of the TEM images of the samples after the reaction, we observed that the cubes had lost their shape and size (Figure 5.16). Therefore, based on the results of conversion and characterization of the catalysts after the reaction, we

Table 5.1 Conversion (micromoles ST) and Percentage Selectivity to ST during Three Consecutive Runs of Gas-Phase Phenylacetylene Hydrogenation

	μmol ST			Selectivity ST (%)		
Sample	*Run 1*	*Run 2*	*Run 3*	*Run 1*	*Run 2*	*Run 3*
Cubes/SiO_2	0.016	0.024	0.019	100	100	100
SiO_2	0.032	0.003	0.002	86	68	64
Ag/SiO_2 imp.	0.510	0.151	0.135	54	80	82
Cubes/TiO_2	0.037	0.043	0.047	92	94	92
TiO_2	0.002	0.003	0.006	82	82	84
Ag/TiO_2 imp.	0.137	0.104	0.112	54–98	85	92
Cubes/Al_2O_3	0.127	0.051	0.016	86	95	95
Al_2O_3	0.030	0.022	0.016	91	91	91
Ag/Al_2O_3 imp.	1.438	0.273	0.173	3–71	84	85

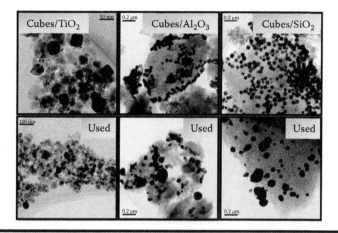

Figure 5.16   TEM images of used Ag catalysts in the gas-phase phenyl-acetylene hydrogenation.

concluded that the samples are not stable in the reaction's conditions. Furthermore, due to the low conversions, we should focus our efforts toward a more favorable system.

5.2.5.2 Liquid-Phase Hydrogenation of PhA

The reactions were performed in steel autoclaves under 3 bar H_2 and 80°C, in toluene. Whether we had a heterogeneous catalyst or the Ag colloids in a water solution (biphasic system), the reactions' conditions were identical. Figure 5.17 shows the evolution of the conversion and selectivity over time for the Ag/Al_2O_3 catalyst series.

The blank experiments (black lines) indicated a conversion around 10% after 4 hours, which increased slightly in time. It is easily observed that once Ag is added to the support, the conversion is increased, and in the case of Ag/Al_2O_3 impregnated catalyst, the reaction goes almost to completion in 4 hours. However, we also noticed that there was no effect of the exposed facet on the selectivity, regardless of the catalyst used. Figures 5.18 and 5.19 present the variation of conversion for the Ag/SiO_2 and Ag/TiO_2 catalysts series, respectively.

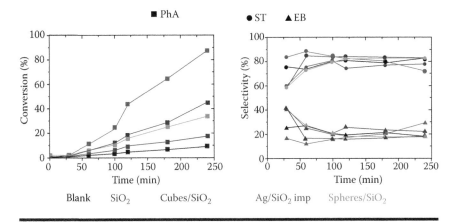

Figure 5.17 **Evolution of the phenylacetylene conversion and styrene and ethylbenzene selectivity in time with Ag/Al$_2$O$_3$.**

In the case of Ag/SiO$_2$ and Ag/TiO$_2$ catalysts, the conversions were lower than in the case of the Ag supported on alumina catalysts, suggesting a support influence. Again, exposed facets of the well-defined catalysts did not influence the selectivity. From the supports tested in this study, alumina produced catalysts that gave the higher conversions. However, based on the results obtained so far, we do not have a plausible explication for these results.

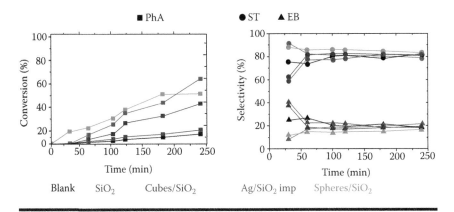

Figure 5.18 **Evolution of the phenylacetylene conversion and styrene and ethylbenzene selectivity in time with Ag/SiO$_2$.**

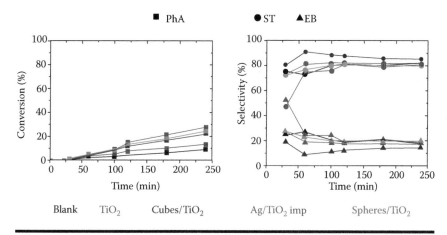

Figure 5.19 Evolution of the phenylacetylene conversion and styrene and ethylbenzene selectivity in time with Ag/TiO₂.

Since the conditions in the gas-phase reactions modified the shape/size of the Ag metal particles in the used catalysts, we also checked all the silver catalysts after the liquid-phase reactions. Figure 5.20 shows the shape of the various supported Ag nanoparticles after reaction. Supporting Ag

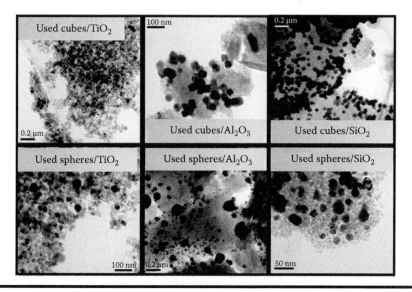

Figure 5.20 TEM images of used Ag catalysts in the liquid phase hydrogenation of phenylacetylene.

nanoparticles led to a strong stabilization of their shape and size (for comparison with the fresh samples, see Figure 5.16). The XRD results (not shown) of the used catalysts also indicated that, after use, the size remained unchanged in comparison with the fresh samples.

Since the kinetic results indicated that the support had an influence on the conversion and selectivity, we also performed the reactions with the pure Ag colloids (cubes and spheres). The results are presented in Figure 5.21.

In the first 4 hours of reaction, the conversions of the two types of colloids follow the same trend as the ones for the supported Ag nanoparticles. However, over time, the conversion increase slows down, which could be attributed to the aggregation of the Ag nanoparticles in solution. Since one of the aims of this project was to determine if the exposed facet had an influence on the selectivity, the results presented in Figure 5.21 show for the first time a 20% difference in the selectivity between the spheres (mainly {111}, 60%) and cubes (mainly {100}, 80%). The drawback of such an approach (reaction with colloids) is that the particles are not as stable as the supported ones. However, we believe that this can be improved by the choice of solvent, temperature, and pressure.

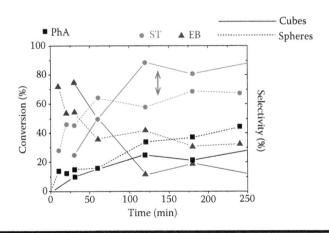

Figure 5.21 **Variation of conversion and selectivity as a function of time for the cube and sphere colloids.**

To summarize, the heterogeneous silver catalysts were found active in both gas- and liquid-phase chemoselective hydrogenation of phenylacetylene. The liquid-phase reactions with the colloids revealed a noticeable influence of the exposed silver facets on the selectivity (cubes vs. spheres). These are rather preliminary results but are promising since silver was only six times less active than palladium, while keeping selectivity on the desirable high values, which is not attainable with palladium.

5.3 Preparation of Silver Nanoparticles

Generally, the catalytic materials can take various forms and their preparation comprises different procedures that involve a multitude of preparation schemes. Moreover, the preparation of a catalyst involves a sequence of several complex processes; many of them can introduce subtle changes in the final material, which can result in dramatic properties/activities of the catalyst. Therefore, it is important that the authors include in their publications the detailed preparation methods used in producing their catalytic materials.

In the following, we will list the procedures used to prepare Ag or Ag-promoted catalysts, mainly heterogeneous, that showed interesting catalytic activities in the examples stated earlier, but also new emerging synthesis used to generate materials with controlled morphologies that, in the future, we are sure will find their way in the mainstream of practical applications.

5.3.1 Synthesis of Well-Defined Ag Nanoparticles

Special emphasis has been given to the preparation methods that have a high degree of control over the cluster size, shape, and dispersion; are reproducible; have a low cost; are suitable to be scaled up; and produce materials in high quantities.

Additionally, the methods should also produce materials that are highly robust (preserve their initial morphologies, or at least the size distribution) under reaction conditions (i.e., contact to reactants, temperatures, pressures, etc.) and suffer minimal coarsening. Since the size and the shape of Ag nano-structures define and control their unique plasmonic proper-ties, there have been many studies on the synthesis of such silver nanostructures for plasmonic applications.

Shape-selective synthesis of Ag nanoparticles has resulted in many shapes (e.g., spheres [67,68], cubes, polyhedral [69], triangular bipyramids [70], prisms [71], and bars, rods, and wires [72]), using a plethora of methods (i.e., reduction with H_2, polyol, photochemical, plasmon-mediated synthesis, etc.) [73]. From these methods, the most promising route—the one with the most recent developments for the synthesis of Ag nanocrystals with controlled shapes—is *the facet-specific capping.* It has been initially the key to the precise control of such particles, and the concept has its origin in heterogeneous catalysis, where chemisorption of atomic or molecular spe-cies from the gas phase often results in radical changes to the shape or morphology of metal nanoparticles [74]. Furthermore, this method seems to address most of the challenges, described before, for such a synthesis to be applicable in catalysis. However, despite the promising data on the synthesis routes, further work should be done in relation to the develop-ment of nonaggressive approaches for the complete removal of undesired encapsulating ligands.

In the literature there are numerous good-quality papers and reviews on the shape-controlled synthesis of silver nano-crystals [75]. Therefore, we are not going to go into detail about the specific procedures, rather, we will emphasize the most important parameters affecting such syntheses, as well as the role of a capping agent in producing a nanoparticle with a specific shape and size.

The first step in a chemical reduction process is the reduc-tion of a silver salt by a reducing agent. The most common

silver precursor is $AgNO_3$. However, recent precursors, such as CF_3COOAg, have also been employed. The advantages associated with the use of this precursor are the lower reaction temperature, elimination of NO_2 derived from the decomposition of NO_3^-, insensitivity to trace impurities from the solvent (especially ethylene glycol), and an easily scaled up process [76]. The chemical reducing agents also play an important role. Typically, for the reduction of dissolved Ag^+ ions to Ag atoms, agents such as sodium borohydride, alcohols, and sodium citrate, as well as ethylene glycol (or products derived from this as a result of reaction conditions, i.e., temperature), are employed. In the second step, neutral silver atoms collide with each other, forming stable nuclei, and growth will take place until all metal ions are consumed. In the final step polymers and surfactans (i.e., capping agents) are used to stabilize the nanostructures, but they can also have a role in directing the particles' growth to the desired shapes.

The success of the **capping agents** (self-assembled mono-layers, surfactants, polymers, and dendrimers) is due to different affinities of the ligands to the exposed crystal faces. A capping agent controls the morphology of a nanocrystal thermodynamically because it reduces the surface free energy (γ) of a specific type of facet through chemisorption or selective binding. For a reaction without a capping agent, under thermodynamic control, the growth of the nanocrystal will progress along the path that minimizes the total surface free energy of the system. In the case of Ag, in the absence of capping agents, the surface free energies of low-index facets increase in the order: $\gamma\{111\} < \gamma\{100\} < \gamma\{110\}$. For example, one can obtain truncated cubes, cuboctahedrons, or octahedrons [77,78] when using Ag nanocubes (enclosed by $\{100\}$) by the gradual replacement of these facets with the more stable $\{111\}$ facets.

Two of the most used capping agents in the synthesis of well-defined Ag nanoparticles are poly(vinyl pyrrolidone) (PVP) and citrate. Sun et al. [79] have shown that PVP can selectively bind to Ag(100) to make its surface free energy lower than that of Ag(111), resulting in the selective formation of nanowires (enclosed by the less stable {100} facets). Nanocubes and nanobars can be obtained with the use of PVP when the reaction conditions are slightly changed [80]. On the other hand, Jin et al. [71] have shown that citrate binds more strongly to Ag(111) than Ag(100), favoring the formation of nanoparticles with {111} facets on the surface, such as prisms.

Computational calculations have added another degree of understanding to the interaction between the capping agent and the Ag surface at the atomic scale. Using ab initio calculations based on the density functional theory, Kilin, Prezhdo, and Xia [81] indicated that the binding energies of the citrate group to Ag(111) and Ag(100) were 13.84 and 3.69 kcal/mol, which would correspond to a difference of six orders in magnitude for the binding constants to Ag(111) and Ag(100) surfaces. The authors concluded that the preferential binding is predetermined by the following factors: (1) coincidence of the symmetries of the ligand and the surface, (2) matching in the size of the ligand and the surface lattice constant, and (3) activation of the ligand electronic structure by hydrogen atom migration. The binding of citric acid to the silver surfaces proceeds through the two methylene-carboxyl groups. Good symmetry and geometry agreement between the acid and the (111) surface produces four ligand surface bonds, whereas only two bonds are able to form with the (100) surface (Figures 5.22 and 5.23).

Al-Saidi, Feng, and Fichthorn [82] used DFT to resolve the role of PVP in the shape-selective synthesis of Ag nanostructures. The authors confirmed that PVP binds more strongly to the Ag(100) rather than to Ag(111), but does not depend on the conformation of the polymer, and that the probability of binding to Ag(100) increased exponentially with the number of repeating units. Various spectroscopic techniques identified that PVP binds to the Ag(100) surface through the oxygen

Citrate   PVP

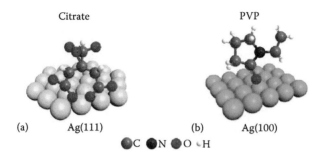

(a)   Ag(111)   (b)   Ag(100)

●C ●N ●O ◦H

Figure 5.22 Schematic illustrations of (a) a citrate group bound to the Ag(111) surface and (b) a repeating unit of PVP bound to the Ag(100) surface. The Ag atoms on the (111) and (100) surfaces are shown as yellow and green spheres, respectively. (Reproduced from X. Xia, J. Zeng, Q. Zhang, C. H. Moran, Y. Xia, *J. Phys. Chem. C* 116 (2012) 21647. With permission.)

and possibly the nitrogen atoms in the 2-purrolidone ring (Figure 5.22b) [83].

Other parameters can also influence the morphology, aggregation state, and stability, such as the initial metal salt concentration, reducing agent/metal salt molar ratio, stabilizer/metal ratio, stabilizer concentration, or other additives (inorganic species). For example, an increase in the molar ratio of reducer to metal salt causes a rapid formation of a large number of nuclei and leads to smaller, monodispersed metal nanoparticles. On the other hand, decreasing the molar ratio leads to a slow formation of a smaller number of nuclei, and results in larger nanoparticles with a greater heterogeneity in size [84].

Other additives (i.e., inorganic species) can also control the shape of the particles. Caswell, Bender, and Murphy [85] and Pilenip [86] were among the first groups to promote the beneficial effect of ionic species (e.g., NaOH) as controlling agents in various types of syntheses and as providers of an alternative to the organic surfactants. The exact mechanism of how these ions worked in the syntheses was not known, and it was concluded that more data were necessary in order to elucidate

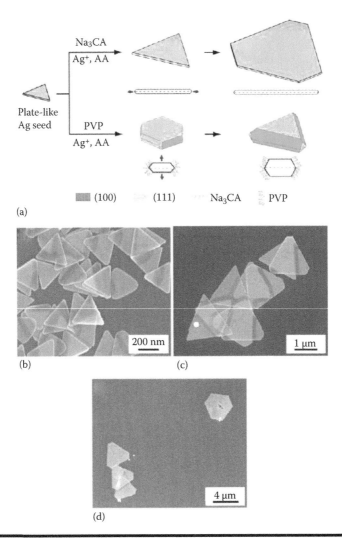

Figure 5.23 (a) Schematic showing the growth of a plate-like Ag seed in the presence of citrate or PVP as the capping agent. For the cross-sectional side views, red arrows mark the major directions of growth. Dashed lines indicate the stacking faults. (b–d) SEM images of thin Ag plates obtained by repeating the seeded growth two, six, and eight times in the presence of citrate as the capping agent. Note that the length scale is considerably increased from (b) to (d).

(Continued)

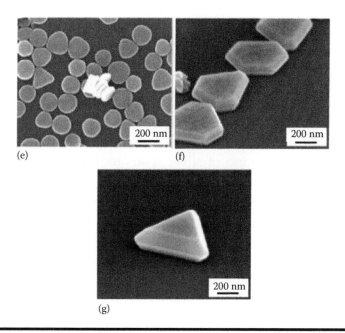

(e)

(f)

(g)

Figure 5.23 (Continued) **(e–g) SEM images of Ag plates obtained by repeating the seeded growth two, six, and eight times in the presence of PVP as the capping agent. (Reproduced from X. Xia, J. Zeng, Q. Zhang, C. H. Moran, Y. Xia, *J. Phys. Chem. C* 116 (2012) 21647. With permission.)**

the ions' effect. Xia's group (University of Washington), well known for the synthesis of well-defined Ag nanostructures, has explored the role(s) of a series of ionic species (e.g., trace amounts of NaCl, Na_2S, or NaHS) in the polyol synthesis [87]. It was discovered that even in trace amounts the ions had a dramatic effect on the synthetic pathways, as well as on the morphologies of both nuclei and products.

The role of chlorine was to enhance oxidation and to etch preferentially twinned particles, leaving only the single-crystal nanoparticles (or seeds) to grow. By controlling the reaction time, single-crystal nanoparticles monodisperse in size, with diameters ranging from 20 to 80 nm, and a mixture of cubes and tetrahedrons was obtained. Chen et al. [88] revealed that the presence of Na_2S also has an important effect on the final morphology. The anion (S_2^-) can combine with silver ions to form Ag_2S colloids. However, an interesting dependency in concentration was discovered: At low concentrations of Na_2S in solution (62.5–250 µM), the ions act as seed catalysts and lead to the rapid formation of silver seeds (final products are nanocubes), whereas at higher concentrations, Na_2S (750 µM) acts as a controlling agent, resulting in reduction of the concentration of free Ag^+ ions in the initial stages, and facilitates the high-yield formation of silver seeds required to form final products as wires.

Based on the literature provided, the synthesis of well-defined nanoparticles is highly dependent in many, if not all the parameters involved and, though seldom reported in the literature, the reproducibility of many chemical methods is greatly affected by trace amounts of contaminants (e.g., solvent contamination) or instrumentation used. One example where the quality of the solvents and the instrumentation used is described in detail is by Skrabalk et al. [89]. By providing all the details, the authors made sure that the entire scientific community is able to follow the synthesis and obtain the same quality of materials. For the future, especially in the light of eventually desired mass production, less toxic and more facile and cost-effective synthetic strategies would be more desirable to develop.

References

1. W. A. Oddy, An unsuspected danger in display, *Museum J.* 73 (1973) 27–28.
2. J. Greeley, M. Mavrikakis, Surface and subsurface hydrogen: Adsorption properties on transition metals and near-surface alloys, *J. Phys. Chem. B* 109 (2005) 3460.
3. C. Christmann, Hydrogen sorption on pure metal surfaces. In *Hydrogen effects in catalysis, fundamentals and practical applications*, Z. Paál, P. G. Menon, (eds.) Marcel Dekker, New York (1988) 12.

4. P. Claus, Selective hydrogenation of α,β-unsaturated aldehydes and other C=O and C=C bonds containing compounds, *Top. Catal.* 5 (1998) 51.

5. A. Borodziński, G. C. Bond, Selective hydrogenation of ethyne in ethene-rich streams on palladium catalysts, Part 2: Steady-state kinetics and effects of palladium particle size, carbon monoxide, and promoters, *Catal. Rev. Sci. Eng.* 50 (2008) 379.

6. A. Molńar, A. Sárkány, M. Varga, Hydrogenation of carbon–carbon multiple bonds: Chemo-, regio- and stereo-selectivity, *J. Mol. Catal. A* 173 (2001) 185.

7. D. Mei, M. Neurock, C. M. Smith, Hydrogenation of acetylene–ethylene mixtures over Pd and Pd–Ag alloys: First-principles-based kinetic Monte Carlo simulations, *J. Catal.* 268 (2009) 181.

8. D. Duca, F. Frusteri, A. Parmaliana, A. G. Deganello, Selective hydrogenation of acetylene in ethylene feedstocks on Pd catalysts, *Appl. Catal. A* 146 (1996) 269.

9. B. Hammer, J. K. Nørskov, Why gold is the noblest of all the metals, *Nature* 376 (1995) 238.

10. J. Sá, G. Tagliabue, P. Friedli, J. Szlachetko, M. H. Rittmann-Frank, F. G. Santomauro, C. J. Milne, H. Sigg, Direct observation of charge separation on Au localized surface plasmons, *Energy Environ. Sci.* 6 (2013) 3584.

11. Y. Nagase, H. Muramatu, T. Sato, Selective hydrogenation of crotonaldehyde to crotyl alcohol on Ag-MnO$_2$/Al$_2$O$_3$·5AlPO$_4$ catalysts, *Chem. Lett.* 17 (1988) 1695.

12. Y. Nagase, H. Nakamura, Y. Yazawa, T. Imamoto, Liquid-phase hydrogenation of crotonaldehyde over silver-manganese oxide catalyst, *Chem. Lett.* 21 (1992) 927.

13. J. L. Solomon, R. J. Madix, Kinetics and mechanism of the oxidation of allyl alcohol on silver(110), *J. Phys. Chem.* 91 (1987) 6241.

14. P. Claus, P. Kraak, R. Schödel, Selective hydrogenation of α,β-unsaturated aldehydes to allylic alcohols over supported mono-metallic and bimetallic Ag catalysts, *Stud. Surf. Sci. Catal.* 108 (1997) 281.

15. P. Claus, P. A. Crozier, P. Druska, Characterization and application of supported metal catalysts with well-tailored pore systems and metal dispersions, *Fresenius J. Anal. Chem.* 361 (1998) 677.

16. P. Claus, H. Hofmeister, Electron microscopy and catalytic study of silver catalysts: Structure sensitivity of the hydrogenation of crotonaldehyde, *J. Phys. Chem. B* 103 (1999) 2766.

17. M. Bron, E. Kondratenko, A. Trunschke, P. Claus, Towards the "pressure and materials gap": Hydrogenation of acrolein using silver catalysts, *Z. Phys. Chem.* 218 (2004) 405.

18. M. Bron, D. Teschner, A. Knop-Gericke, B. Steinhauer, A. Scheybal, M. Hävecker, D. Wang, R. Födisch, D. Hönicke, A. Wootsch, R. Schlögl, P. Claus, Bridging the pressure and materials gap: In-depth characterisation and reaction studies of silver-catalysed acrolein hydrogenation, *J. Catal.* 234 (2005) 37.

19. M. Bron, D. Teschner, A. Knop-Gericke, A. Scheybal, B. Steinhauer, M. Hävecker, R. Födisch, D. Hönicke, R. Schlögl, P. Claus, In situ-XAS and catalytic study of acrolein hydrogenation over silver catalyst: Control of intramolecular selectivity by the pressure, *Catal. Commun.* 6 (2005) 371.

20. M. Bron, D. Teschner, A. Knop-Gericke, F. C. Jentoft, J. Kröhnert, J. Hohmeyer, C. Volckmar, B. Steinhauer, R. Schlögl, P. Claus, Silver as acrolein hydrogenation catalyst: Intricate effects of catalyst nature and reactant partial pressures, *Phys. Chem. Chem. Phys.* 9 (2007) 3559.

21. W. Grünert, A. Brückner, H. Hofmeister, P. Claus, Structural properties of Ag/TiO_2 catalysts for acrolein hydrogenation, *J. Phys. Chem. B* 108 (2004) 5709.

22. C. Hamel, M. Bron, P. Claus, A. Seidel-Morgenstern, Experimental and model based study of the hydrogenation of acrolein to ally alcohol in fixed bed and membrane reactors, *Int. J. Chem. React. Eng.* 3 (2005) A10.

23. M. Bron, D. Teschner, U. Wild, B. Steinhauer, A. Knop-Gericke, C. Volckmar, A. Wootsch, R. Schlögl, P. Claus, Oxygen-induced activation of silica supported silver in acrolein hydrogenation, *Appl. Catal. A* 341 (2008) 127.

24. M. Steffan, A. Jakob, P. Caus, H. Lang, Silica supported silver nanoparticles from a silver(I) carboxylate: Highly active catalyst for regioselective hydrogenation, *Catal. Commun.* 10 (2009) 437.

25. R. Van Hardeveld, F. Hartog, The statistics of surface atoms and surface sites on metal crystals, *Surf. Sci.* 15 (1969) 189.

26. C. E. Volckmar, M. Bron, U. Bentrup, A. Martin, P. Claus, Influence of the support composition on the hydrogenation of acrolein over $Ag/SiO_2–Al_2O_3$ catalysts, *J. Catal.* 261 (2009) 1.

27. J. E. Bailie, G. J. Huntchings, Promotion by sulfur of Ag/ZnO catalysts for the hydrogenation of but-2-enal, *Catal. Commun.* 2 (2001) 291.

28. Y. Önal, M. Lucas, P. Claus, Application of a capillary microreactor for selective hydrogenation of α,β-unsaturated aldehydes in aqueous multiphase catalysis, *Chem. Eng. Technol.* 28 (2005) 972.

29. M. Steffan, M. Lucas, A. Brandner, M. Wollny, N. Oldenburg, P. Claus, Selective hydrogenation of citral in an organic solvent, in a ionic liquid, and in substance, *Chem. Eng. Technol.* 30 (2007) 481.

30. F. Haass, M. Bron, H. Fuess, P. Claus, In situ X-ray investigations on AgIn/SiO$_2$ hydrogenation catalysts, *Appl. Catal. A* 318 (2007) 9.

31. K. Brandt, M. E. Chui, D. J. Watson, M. S. Tikhov, R. M. Lambert, Chemoselective catalytic hydrogenation of acrolein on Ag(111): Effect of molecular orientation on reaction selectivity, *J. Am. Chem. Soc.* 131 (2009) 17286.

32. B. C. Khanra, Y. Jugnet, J. C. Bertolini, Energetics of acrolein hydrogenation on Pt(1 1 1) and Ag(1 1 1) surfaces: A BOC-MP model study, *J. Mol. Catal. A* 208 (2004) 167.

33. K. H. Lim, A. B. Mohammad, I. V. Yudanov, K. M. Neyman, M. Bron, P. Claus, N. Rösch, Mechanism of selective hydrogenation of α,β-unsaturated aldehydes on silver catalysts: A Density Functional Study, *J. Phys. Chem. C* 113 (2009) 13231.

34. X. Yang, A. Wang, X. Wang, T. Zhang, K. Han, J. Li, Combined experimental and theoretical investigation on the selectivities of Ag, Au, and Pt catalysts for hydrogenation of crotonaldehyde, *J. Phys. Chem. C* 113 (2009) 20918.

35. Y. Chen, C. Wang, H. Liu, J. Qiu, X. Bao, Ag/SiO$_2$: A novel catalyst with high activity and selectivity for hydrogenation of chloronitrobenzenes, *Chem. Commun.* (2005) 5298.

36. B. Li, H. Li, Z. Xu, Experimental evidence for the interface interaction in Ag/C$_{60}$ nanocomposite catalyst and Its crucial influence on catalytic performance, *J. Phys. Chem. C* 113 (2009) 21526.

37. H. Zhang, X. Li, G. Chen, Ionic liquid-facilitated synthesis and catalytic activity of highly dispersed Ag nanoclusters supported on TiO$_2$, *J. Mater. Chem.* 19 (2009) 8223.

38. K.-I. Shimizu, Y. Miyamoto, A. Satsuma, Size- and support-dependent silver cluster catalysis for chemoselective hydrogenation of nitroaromatics, *J. Catal.* 270 (2010) 86.

39. K. Shimizu, Y. Miyamoto, T. Kawasaki, T. Tanji, Y. Tai, S. Satsuma, Chemoselective hydrogenation of nitroaromatics by supported gold catalysts: Mechanistic reasons of size- and support-dependent activity and selectivity, *J. Phys. Chem. C* 113 (2009) 17803.
40. A. Corma, P. Serna, Chemoselective hydrogenation of nitro compounds with supported gold catalysts, *Science* 313 (2006) 332.
41. A. Leelavathi, T. U. B. Rao, T. Pradeep, Supported quantum clusters of silver as enhanced catalysts for reduction, *Nanoscale Res. Lett.* 6 (2011) 123.
42. R. Campostrini, G. Carturan, R. M. Baraka, Propene hydrogenation with Pt/Ag intermetallic phases supported on SiO_2 gels, *J. Mol. Catal.* 78 (1993) 169.
43. H. Imamura, K. Fujita, Y. Sakata, S. Tsuchiya, Hydrogenation properties of lanthanide-dosed Ag bimetallic catalysts, *Catal. Today* 28 (1996) 231.
44. H. Ehwald, A. A. Shestov, V. Muzykantov, Ethylene hydrogenation mechanism on Ag/SiO_2 catalysts elucidated by isotope kinetics, *Catal. Lett.* 25 (1994) 149.
45. A. Sárkány, Zs. Révay, Some features of acetylene and 1,3-butadiene hydrogenation on Ag/SiO_2 and Ag/TiO_2 catalysts, *Appl. Catal. A* 243 (2003) 347.
46. M. Balooch, M. J. Cardillo, D. R. Miller, R. E. Stickney, Molecular beam study of the apparent activation barrier associated with adsorption and desorption of hydrogen on copper, *Surf. Sci.* 46 (1974) 358.
47. V. Amir-Ebrahimi, J. J. Rooney, Organic Molecular Probes in heterogeneous catalysis. Hydrogenation of norbornadiene on gold, *Catal. Lett.* 127 (2009) 20.
48. A. B. Mohammad, K. H. Lim, I. V. Yudanov, K. M. Neyman, N. Rösch, A computational study of H_2 dissociation on silver surfaces: The effect of oxygen in the added row structure of Ag(110), *Phys. Chem. Chem. Phys.* 9 (2007) 1247.
49. E. Yagasaki, R. I. Masel, Specialist periodical reports: Catalysis, *R. Soc. Chem. Lond.* 11 (1994) 164.
50. R. B. Moyes, P. B. Wells, J. Grant, N. Y. Salman, Electronic effects in butadiene hydrogenation catalysed by the transition metals, *Appl. Catal. A* 229 (2002) 251.
51. E. Nishimura, Y. Inoue, I. Yasumori, The mechanism of the selective hydrogenation of 1,3-butadiene on copper surfaces, *Bull. Chem. Soc. Jpn.* 48 (1975) 803.

52. G. A. Somorjai, C. Aliaga, Molecular studies of model surfaces of metals from single crystals to nanoparticles under catalytic reaction conditions. Evolution from prenatal and postmortem studies of catalysts, *Langmuir* 26 (2010) 16190.

53. J. Singh, E. M. C. Alayon, M. Tromp, O. V. Safonova, P. Glatzel, M. Nachtegaal, R. Frahm, J. A. van Bokhoven, Generating highly active partially oxidized platinum during oxidation of carbon monoxide over Pt/Al_2O_3: In situ, time-resolved, and high-energy-resolution X-ray absorption spectroscopy, *Angew. Chem. Int. Ed.* 47 (2008) 9260.

54. J. A. van Bokhoven, C. Louis, J. T. Miller, M. Tromp, O. V. Safonova, P. Glatzel, Activation of oxygen on gold/alumina catalysts: In situ high-energy-resolution fluorescence and time-resolved X-ray spectroscopy, *Angew. Chem. Int. Ed.* 45 (2006) 4651.

55. J. A. McCaulley, In-situ x-ray absorption spectroscopy studies of hydride and carbide formation in supported palladium catalysts, *J. Phys. Chem.* 97 (1993) 10372.

56. N. A. Zaitseva, V. V. Molchanov, V. V. Chesnokov, R. A. Buyanov, V. I. Zaikovskii, Catalysts based on filamentous carbon in the hydrogenation of aromatic compounds, *Kinet. Catal.* 44 (2003) 129.

57. V. I. Zaikovskii, V. V. Chesnokov, R. A. Buyanov, Catalyst and technology for production of carbon nanotubes, *Kinet. Catal.* 43 (2002) 677.

58. J. Silvestre-Albero, M. Borasio, G. Rupprechter, H.-J. Freund, Combined UHV and ambient pressure studies of 1,3-butadiene adsorption and reaction on Pd(1 1 1) by GC, IRAS and XPS, *Catal. Commun.* 8 (2007) 292.

59. J. C. Chen, B. D. DeVries, B. Fruhberger, C. M. Kim, Z. M. Liu, Spectroscopic characterization of thin vanadium carbide films on a vanadium (110) surface: Formation, stability and reactivities, *J. Vac. Sci. Tech. A* 13 (1995) 1600.

60. M. Kreich, P. Claus, Direct conversion of linoleic acid over silver catalysts in the presence of H_2: An unusual way towards conjugated linoleic acids, *Angew. Chem. Int. Ed.* 44 (2005) 7800.

61. S. Sugawa, K. Sayama, K. Okabe, H. Arakawa, Methanol synthesis from CO_2 and H_2 over silver catalyst, *Energy Convers. Mgmt.* 36 (1995) 665.

62. J. Wambach, A. Baiker, A. Wokaum, CO_2 hydrogenation over metal/zirconia catalysts, *Phys. Chem. Chem. Phys.* 1 (1999) 5071.

63. T. P. Maniecki, P. Mierczynski, W. Maniukiewiz, K. Bawolak, D. Gebauer, W. K. Jozwiak, Bimetallic Au–Cu, Ag–Cu/CrAl$_3$O$_6$ catalysts for methanol synthesis, *Catal. Lett.* 130 (2009) 481.

64. A. R. Vilchis, M. Avalos-Borja, S. A. Gómez, J. A. Hernández, A. Olivas, T. A. Zepeta, Alternative bio-reduction synthesis method for the preparation of Au(AgAu)/SiO$_2$–Al$_2$O$_3$ catalysts: Oxidation and hydrogenation of CO, *Appl. Catal. B* 90 (2009) 64.

65. E. A. Shaw, T. Rayment, A. P. Walker, J. R. Jennings, R. M. Lambert, Methanol synthesis activity of Au/CeO$_2$ catalysts derived from a CeAu$_2$ alloy precursor: Do schottky barriers matter? *J. Catal.* 126 (1990) 219.

66. C. Paun, Second year report to the Swiss National Fund, project no. SNF report project 200021 137731, J. A. van Bokhoven, J. Sá.

67. P. Y. Silvert, R. Herrera-Urbina, N. Duvauchelle, V. Vijayakrishnan, K. T. Elhsissen, Preparation of colloidal silver dispersions by the polyol process. Part 1—Synthesis and characterization, *J. Mater. Chem.* 6 (1996) 573.

68. A. Pyatenko, M. Yamaguchi, M. Suzuki, Synthesis of spherical silver nanoparticles with controllable sizes in aqueous solutions, *J. Phys. Chem. C* 111 (2007) 7910.

69. A. Tao, P. Sinsermsuksakul, P. Yang, Polyhedral silver nanocrystals with distinct scattering signatures, *Angew. Chem. Int. Ed.* 45 (2006) 4597.

70. J. Zhang, S. Li, J. Wu, G. C. Schatz, C. A. Mirkin, Plasmon-mediated synthesis of silver triangular bipyramids, *Angew. Chem. Int. Ed.* 48 (2009) 7787.

71. R. Jin, Y. Cao, C. A. Mirkin, K. L. Kelly, G. C. Schatz, J. G. Zheng, Photoinduced conversion of silver nanospheres to nanoprisms, *Science* 294 (2001) 1901.

72. B. Wiley, Y. Sun, Y. Xia, Synthesis of silver nanostructures with controlled shapes and properties, *Acc. Chem. Res.* 40 (2007) 1067.

73. M. Rycenga, C. M. Cobley, J. Zeng, W. Li, C. H. Moran, Q. Zhang, D. Qin, Y. Xia, Controlling the synthesis and assembly of silver nanostructures for plasmonic applications, *Chem. Rev.* 111 (2011) 3669.

74. Q. Chen, N. V. Richardson, Surface facetting induced by adsorbates, *Progress Surf. Sci.* 73 (2003) 59.

75. X. Xia, J. Zeng, Q. Zhang, C. H. Moran, Y. Xia, Recent developments in shape-controlled synthesis of silver nanocrystals, *J. Phys. Chem. C* 116 (2012) 21647.

76. Q. Zhang, W. Li, L.-P. Wen, J. Chen, Y. Xia, Facile Synthesis of Ag nanocubes of 30 to 70 nm in edge length with CF_3COOAg as a precursor, *Chem. Eur. J.* 16 (2010) 10234.
77. A. R. Tao, S. Habas, P. Yang, Shape control of colloidal metal nanocrystals, *Small* 4 (2008) 310.
78. L. Vitos, A. V. Ruban, H. L. Skriver, J. Kollar, The surface energy of metals, *Surf. Sci.* 411 (1998) 186.
79. Y. Sun, B. Mayers, T. Herricks, Y. Xia, Polyol synthesis of uniform silver nanowires: A plausible growth mechanism and the supporting evidence, *Nano Lett.* 3 (2003) 955.
80. B. J. Wiley, Y. Chen, J. M. McLellan, Y. Xiong, Z. Y. Li, D. Ginger, Y. Xia, Synthesis and optical properties of silver nanobars and nanorice, *Nano Lett.* 7 (2007) 1032.
81. D. S. Kilin, O. V. Prezhdo, Y. Xia, Shape-controlled synthesis of silver nanoparticles: Ab initio study of preferential surface coordination with citric acid, *Chem. Phys. Lett.* 458 (2008) 113.
82. W. A. Al-Saidi, H. Feng, K. A. Fichthorn, Adsorption of Polyvinylpyrrolidone on Ag surfaces: Insight into a structure-directing agent, *Nano Lett.* 12 (2012) 997.
83. C. H. Moran, M. Rycenga, Q. Zhang, Y. Xia, Replacement of poly(vinyl pyrrolidone) by thiols: A systematic study of Ag nanocube functionalization by surface-enhanced Raman Scattering, *J. Phys. Chem. C* 115 (2011) 21852.
84. T. M. Tolaymat, A. M. El Badawy, A. Genaidy, K. G. Scheckel, T. P. Luxtona, M. Suidan, An evidence-based environmental perspective of manufactured silver nanoparticle in syntheses and applications: A systematic review and critical appraisal of peer-reviewed scientific papers, *Sci. Total Environ.* 408 (2010) 999.
85. K. K. Caswell, C. B. Bender, C. J. Murphy, Seedless, surfactantless wet chemical synthesis of silver nanowires, *Nano Lett.* 3 (2003) 667.
86. M. P. Pileni, The role of soft colloidal templates in controlling the size and shape of inorganic nanocrystals, *Nat. Mater.* 2 (2003) 145.
87. B. Wiley, T. Herricks, Y. Sun, Y. Xia, Polyol synthesis of silver nanoparticles: Use of chloride and oxygen to promote the formation of single-crystal, truncated cubes and tetrahedrons, *Nano Lett.* 4 (2004) 1733.
88. D. Chen, X. Qiao, X. Qiu, J. Chen, R. Jiang, Convenient, rapid synthesis of silver nanocubes and nanowires via a microwave-assisted polyol method, *Nanotechnology* 21 (2010) 025607.
89. S. E Skrabalak, L. Au, X. Li, Y. Xia, Facile synthesis of Ag nanocubes and Au nanocages, *Nature Protocols* 7 (2007) 2182.

Index

Page numbers ending in f refer to figures. Page numbers ending in t refer to tables.

Printed and bound by CPI Group (UK) Ltd, Croydon, CR0 4YY

22/10/2024

01777613-0005